OPTIMIZATION METHODS FOR A STAKEHOLDER SOCIETY

Nonconvex Optimization and Its Applications

Volume 73

OPTIMIZATION METHODS FOR A STAKEHOLDER SOCIETY

A Revolution in Economic Thinking by Multi-objective Optimization

By

Willem K. Brauers
Faculty of Applied Economics and
Institute for Development Policy and Management
University of Antwerp
Belgium

Springer Science+Business Media, LLC

Electronic Services <http://www.wkap.nl>

Library of Congress Cataloging-in-Publication

Brauers, Willem K.
Optimization Methods for a Stakeholder Society: A Revolution in Economic Thinking by Multi-objective Optimization
ISBN 978-1-4613-4824-5 ISBN 978-1-4419-9178-2 (eBook)
DOI 10.1007/ 978-1-4419-9178-2

Printed on acid-free paper.

TABLE OF CONTENTS

PART THREE NEW ECONOMICS FOR A STAKEHOLDER SOCIETY

PART FOUR REFERENCE POINT THEORY A METHOD TO COMPARE UNTIL NOW INCOMPARABLE ALTERNATIVES

PART FIVE A MULTIPLICATIVE FORM FOR MULTIOBJECTIVE OPTIMIZATION

PART SIX RATIO ANALYSIS A SEPARATE METHOD FOR MULTIPLE OBJECTIVES OPTIMIZATION?

INTRODUCTION

Storytelling about the "Best Future" for the world is not an easy job. Let us call it a SWOT-analysis. On the one side, there are Weaknesses and Threats, on the other Strengths and Opportunities. Among Weaknesses and Threats we think for instance of global warming of the earth, of its over-population, of the AIDS tragedy and of the antiglobalization backlash versus globalization of the economy.

Among Strengths and Opportunities, we think of the birth of the Eurozone and the rebirth of sciences, under which we include the reform of the theory of economics.

Our goal is to contribute this reform. A central position will be taken, not by one decision-maker dictator, but by all interested members of the stakeholder society. These people however have different aspirations, sometimes in contradiction with each other. Optimization for humanity will be sought, but optimization is also necessary for every person separately. Indeed every person entertains many aspirations, sometimes under the form of objectives independent from each other. We shall take these independent objectives into consideration, striving at the same time for an optimal solution. Is it a revolution in economic thinking? The reader will judge.

WHICH ECONOMICS?

Robbins stated that "economics is neutral between ends"[1], the ends being on the one hand scarce means or resources and on the other the numerous needs, which can be translated into "objectives", both ends being material or non-material.

Consequently, economics cannot be normative. In that case, the objectives themselves interfere in economics. Orwell's "Big Brother" gives his orders by television; ipso facto, the price of television sets has to be zero. Some religions forbid interest payments although it is the price of money. A government may restrict publicity that tries to create new needs, etc.

The human being is stimulated by many aspirations or, more specifically by several objectives, which however cannot be brought under one denominator. However an old dictum for economists states that resources are limited and consequently a problem of choice arises. Moreover in order to compare resources and objectives they have to be measurable, either cardinally or ordinally. Economics is then defined as: a human science, which studies the possibilities of choice in a human society between measurable scarce resources and measurable numerous wants, translated into several objectives. These objectives can be dependent, interdependent or independent. Dependence would mean either a cause-

and-effect phenomenon or a statistical correlation between each other; interdependence that the objectives are mutually influencing each other. Besides dependence and interdependence, the book will mainly tackle the independence of the objectives. Moreover, dependence and interdependence could be instruments to diminish the number of independent objectives.

Welfare economics signify that to the greatest extent possible material objectives have to be fulfilled given a set of material constraints. *Well-being economics* would rather mean that to the greatest extent possible material and immaterial objectives have to be fulfilled given a set of material and immaterial constraints. Marshall says that we have to deal with man as he is; not with an abstract or 'economic man'; but a man of flesh and blood, which means a condemnation of the abstract notion of the so-called "homo economicus"[2]. The only limitation of economics consists in the fact that resources and needs have to be measurable. Even a ranking, a dominance or a hierarchy is sufficient.

The actual world is facing many problems in the field of human behavior and their responses on the environment. The so-called green parties all over the world maintain that solutions have to be found outside economics, as economics is not capable to solve these problems. The point of view taken above about the significance of economics proves that this standpoint of the green parties is wrong. Economics indeed will also tackle the problems of environment, pollution and all immaterial needs. This will be further illustrated in the text.

THE AIM OF THIS WORK

Aspirations, read objectives, of the human being are numerous but limited by the available resources. Aiming at an optimal and global fulfillment of all objectives we agree with Steuer when he writes:

> "Since multiple objective problems rarely have points that simultaneously maximize all of the objectives, we are typically in a situation of trying to maximize each objective to the greatest extent possible"[3].

Optimization is sometimes replaced by categories such as: excellent, good, sufficient or insufficient; no, moderate or strong impact; low, medium or high cost. These categories have to be quantified, either ordinally or cardinally, in order to provide a quantitative optimization[4]. An *Attribute* designates then the quantitative form of an objective[5]. In order to satisfy the objectives two courses are open. On the one hand alternative solutions or projects, we simply call them *Alternatives*, are already present (*Discrete Case*). On the other hand the process of optimization itself generates the alternatives (*Continuous Case*).

What is the importance for practical life? On the one side an industrial firm aims at an increase in productivity by a better motivation of the personnel and by an improvement in the safety of the workers and on the other at an increase in its market share. Additionally more and more, a firm will be pressed to satisfy ecological conditions "to the greatest extent possible". The distribution sector could strive for the maximization of Net Present Value, the minimization of the break-even period in the case of a new firm and to an optimal penetration in the European Union or in world markets.

Project Analysis on the one hand may aim at the maximization of NPV, on the other at the maximization of employment.

Concerning privatization of state enterprises or services the potential buyer will look after a minimum takeover price, a possible payback period, Net Present Value and Internal Rate of Return, whereas the state will look after a maximum takeover bid, employment, value added and new investments[6].

Speaking about "human society" the modern concept of *Stakeholder* has to be introduced. A private company, like the TNT Post Group in the Netherlands, informs all kinds of groups of stakeholders, such as big institutional shareholders, private shareholders, analysts and financial journalists. Argenti maintains that a company should be run for the benefits of all those who may have a “stake” in it, such as investors, employees, customers, suppliers and the relevant community[7].

In a variant of the theory, the Equity version, all stakeholders obtain an equal status, which is against classical capitalism where the return solely goes to the shareholders.

Anyway, in fixing the objectives, one has to take into account all kinds of pressure groups such as trade unions, employer organizations, consumer organizations, groups of ecologists or lobbies of all kinds. Additionally the state considers itself as the promoter of the general interest.

In the past multiobjective optimization acted by a decision-maker-dictator, assisted by an adviser-researcher, who proposed the application of a multiobjective model. Nowadays, the stakeholders, assisted by especially knowledgeable people, take the central position. These especially knowledgeable people are the indirect representatives of the stakeholders by parliament, consumer organizations, trade unions or associations of employers, which on their turn are assisted by study centers. Finally, if there is a principal he will take a final decision.

The book is divided in six parts. In a first part the link is made with traditional economics, whereas the second part shows the advent of new economics. The third part brings the new economics for a stakeholder society with, as much as possible, an optimization of the fulfillment of the aspirations of the individual and of all the stakeholders. However, despite hierarchy, filtering, indifference and dominance, objectives and alternatives could remain incomparable. In order to solve this problem, Reference Point Theory is proposed in the fourth part of the book. In the fifth part another nonlinear approach is presented, namely the Multiplicative

Representation. In the sixth part the question is posited, if still another approach, namely Ratio Analysis, represents a control mechanism on the previous approaches, or can be considered as an additional nonlinear theory.

Each nonlinear theory is illustrated with many examples and applications. The reader who is especially interested in nonlinear methods for optimization with different independent objectives, could immediately start with:

Part IV: Reference Point Theory
Part V: Multiplicative Representation
Part VI: Ratio Analysis for Multiobjective Optimization

with feedback for clarification by the preceding parts.

NOTES INTRODUCTION

1. L.Robbins, *An Essay on the Nature and Significance of Economic Science*, London 1935.

2. A.Marshall, *Principles of Economics*, New York, MacMillan,, 8th ed. 1952, VI.

3. R.E.Steuer, , *Multiple Criteria Optimization : Theory, Computation and Application*, Krieger, Malabar (Fla., U.S.), 1989, 138.

4. W.K.Brauers, *Some Principles of Multi-Optimization*, Systems Sciences and Engineering, Edited by Cheng Wei-Min, International Academic Publishers and Pergamon, Beijing (Ch.), 1988, 678-682.

5 Keeney and Raiffa present the example of the objective: reduce sulfur, dioxide emissions, which is measured by the attribute: tons of sulfur dioxide emitted per year. R.L.Keeney, H.Raiffa, *Decisions with Multiple Objectives. Preferences and Value Tradeoffs*, Cambridge University Press, USA, 1993, 32.

6 W.K.Brauers, *PRIVATA: A model for privatization with multiple non-transitive objectives*, Public Choice, Vol. 85, N° 3-4, 1995, 353-370,

7. J.Argenti, *Stakeholders, the case against*, Long Range Planning, vol. 30, N° 3, 442. See also: A.Campbell, *Stakeholders, the case in favour*, Long Range Planning, 1997 vol. 30, N° 3, 446-449 and Long Range Planning, 1998 vol. 31, N° 2 with as special theme: *the Stakeholder Corporation.*

PART ONE
THE LINK WITH MORE TRADITIONAL ECONOMICS

CHAPTER 1

NEW ECONOMICS AGAINST THE HISTORY OF ECONOMIC THOUGHT

As already mentioned in the introduction, the objectives of the individual are important building blocks for a stakeholder society. Already the classical economists had an open view on this issue. General interdependence considerations for a human society come next. Finally, the introduction of mathematics into economics will form another important starting point. A narrower view of economics, such as normative economics, will not be treated in this book.

1.1 CLASSICAL ECONOMISTS ABOUT OBJECTIVES

The classical authors do not speak of "objectives", but of means of subsistence (Malthus, 1766-1834[1]), wants (Senior, 1790-1864[2]; Marshall, 1842-1924[3]), pleasure (Gossen, 1810-1859[4]; Jevons, 1835-1882[5]). However, they also speak of limitations in the means of subsistence (Malthus), scarcity (Senior) or pain (Gossen, Jevons).

Already in his time Gossen stated: "Enjoyment must be so arranged that the total life pleasure should become a maximum"[6]. Guitton summarized the classical attitude very well when he wrote that economic man in a rational way attempts to minimize his pains and maximize his pleasures[7]. For A. Smith (1723-1790) the optimum for human society will be reached by the liberty of the individual and with free market competition, whereas the state will only play a complementary role[8]. This "natural order" and the strength of the master versus labor, which was treated as merchandise, would create opposing reactions. Indeed the reaction of K. Marx (1818-1883), with his view on human society, protecting the lower classes in the struggle of the classes by the common ownership of the means of production, presented for him an optimal solution[9].

Malthus introduced into economic thought a form of mathematics through his formula of geometrical progression of population against the arithmetical progression of subsistence[10].

1.2 GENERAL ECONOMIC INTERDEPENDENCE

F. Quesnay (1694-1774) was the forerunner of General Economic Interdependence through his famous "Tableau Economique" (1758). Being the personal physician of Louis XV of France, he was inspired by the discovery of the circulation of the blood in the human body by William Harvey (1578-1657)[11].

The mathematical school, represented by L. Walras (1834-1910), stressed the general interdependence of economic equilibrium in human society. The work of Walras was later continued by W. W. Leontief (1906-1999, Nobel prize 1973) with his Input-Output Analysis. Leontief brought a more empirical approach to general economic interdependence than Walras. Leontief introduced empirical aggregation and inversion of the variables in the model.

1.3 THE PARTIAL APPROACH

Contrary to the general approach, a tendency originated in economics to a more partial approach, sometimes by means of abstraction. The 'ceteris paribus principle' and the 'homo oeconomicus' represent typical examples of this approach. They made economic reasoning much easier. Ceteris paribus means that only one variable is changed while all other variables are kept unchanged. The homo oeconomicus is an economic abstraction of the human being, who would only react to economic laws and formulas.

A.Marshall stands for partial equilibrium. Instead of connecting interdependently all variables of an economic system, some variables are studied separately in a step by step approach.

> "By that method we fix our minds on some central points: we suppose it for the time to be reduced to a stationary state; and we then study in relation to it the forces that affect the things by which it is surrounded, and any tendency there may be to equilibrium of these forces. A number of these partial studies may lead the way towards a solution of problems too difficult to be grasped at one effort"[12].

Although J. M. Keynes (1883-1946) formulates some criticisms of Marshall, he remains faithful to his teacher and he follows Marshall by making partial studies[13]. Instead of making a study of general economic equilibrium, Keynes limits himself to a study of partial equilibrium and then only limited to partial equilibrium for the short period[14].

It is beyond doubt that Keynes, due to his mathematical schooling in Eton and Cambridge and familiar with economic sciences thanks to his father John Neville Keynes, professor at Cambridge, knew the works of Walras[15]. Whereas Keynes very extensively cited the works of the Mercantilists, the Physiocrats and what he named the Classical School, in his General Theory he only mentioned Walras once and then only in relation with the interest rate[16]. He gave the impression of having no high estimation either for Walras or for general economic equilibrium. Indeed his biographer, Sir Harrod, was of that opinion:

> "I should suppose that Keynes, whether he read Walras or not, was of the opinion that he was not of great significance, thereby following Marshall. I should think that the Cambridge view at the time was that Walras was using very inelegant mathematics to demonstrate what was well known already, viz. the general theory of mutually inter dependent equilibrium"[17].

The partial approach is also observable in the many subdivisions which are made in economics. For example, concerning the scission between *Statics* and *Dynamics*, many rather diverging definitions exist. Samuelson-Frisch call a system dynamical if functional equations with variables at different points of time are essentially involved[18]. Every other analysis of economics belongs then to statics. This is the case for instance with the *Cinematic Model* of Guitton in which the variable 'time' is a 'catch all' variable covering all our ignorance about remaining data[19].

It is true that the Samuelson-Frisch interpretation already belongs to mathematical economics. With the reigning confusion nobody can say it better than Samuelson, when he maintains that often in the writing of economists, dynamic and static are simply synonyms for good and bad, realistic and unrealistic, simple and complex. One's own theory is dynamic, those of others static[20].

Indeed for instance J. M. Clark identifies dynamic studies as being imbued with realism and empiricism[21].

Anyway, other definitions seem to be confused and the definition of Hicks does not withstand the definition of Samuelson-Frisch. Hicks speaks of Economic Statics, if one does not consider dating, Economic Dynamics where every quantity is dated. He cites the example of an entrepreneur who is not asked when the production factors are employed or when the products are ready. In economic dynamics special attention is given to the effects of changes in the relations between factors and products[22]. This definition of dynamics is certainly too broad and covers some items which certainly belong to statics, like a dated photograph of a quantity of products, whereas the comparison of two photographs of a quantity of products which are taken at another point of time belongs to *Comparative Statics*. Hicks is however right that there is not much in common with the distinction between dynamics and statics in the physical sciences[23]. In common parlance dynamic could also mean movement. The same point of view is reflected by certain economists like J. B. Clark when he states that dynamics, in contrast to statics, assumes changes in the basic data of the economy[24]. Others like J. Tinbergen will rather call it *Foundations* and *Structure* which "as a rule, change gradually or infrequently[25]. On the contrary, economic policy and also cyclical movements will rather stress change.

Finally we mention the difference between micro- and macroeconomics. *Macroeconomics* demonstrates the relationships between the main production branches, other available resources, consumption, investments, savings and moneyflows of the regional, national and international economies. *Microeconomics* presents the study of the market mechanism, the household economy and the firm as brought by authors like A. Marshall, E. H. Chamberlin, J. Robinson and others, beside works on management science and operations research[26].

J. Tinbergen put the emphasis for macroeconomics on the abstraction side in comparison with microeconomics[27]. Others see the difference rather in the degree of aggregation[28].

We define *Aggregation* as the summation of not entirely homogeneous items to functional totals. Walras does not propose aggregation but Leontief does. He adds non-homogenous, very detailed production factors, to aggregated production factors of a production branch, and non-homogeneous products to an aggregated production branch. Every time, he uses prices as weighting factors. In his 'General Theory' of 1936 Keynes speaks of consumption, investments and savings, but it is only during the war, when he is engaged in the national bookkeeping of England, that he applies the corresponding but ex-post aggregations[29].

One of the failures of Soviet planning originated from non-aggregation of the hundreds of thousands of material balances at a lower level to some thousands of material balances of the Gosplan at the higher level. Moreover any link was missing with the intersectoral balances in value terms at the national level. Also a market economy would have troubles in summing up production functions from a non-aggregated bottom up approach. On the other side a top down approach like a Cobb-Douglas production function fails to have a link with production functions at a lower level.

Is aggregation also not limited through the independence of the objectives of the stakeholders? This point has to be clarified in detail later.

1.4 THE RISE OF MATHEMATICS IN ECONOMICS

In more recent years the power of the mathematical method gained more recognition. With Walras we already saw the use of mathematics in economics or shall we rather call it *Mathematical Economics*[30]? In the nineteen thirties a new tendency originated: the upswing of Econometrics.

Is *Econometrics* a marginal science or a science of synthesis? J. Tinbergen, one of the first pioneers of econometrics, defines econometrics as a marginal field between mathematical economics and mathematical statistics[31]. R. Frisch, another pioneer of econometrics, rather defines it as "the advancement of economic theory in its relation to statistics and mathematics"[32]. After the definition of van Waterschoot et al.:

> "econometrics is essentially a science of synthesis of economic theory and empirical observation with the help and use of statistics and mathematics"[33],

we may conclude that econometrics is a science which studies the economy in an explicit mathematical formulation but at the same time using statistical data and methods. It means that statistical data are used but also statistical inference and even statistical methods in the largest sense, such as correlation, regression analysis etc.[34].

The Econometric Society was founded in 1933 as "an International Society for the advancement of economic theory in its relation to statistics and mathematics", subtitle of 'Econometrica' the journal of the Society since 1933. The persons who took the initiative for the foundation of the Society were R. Frisch and F. Divisia[35].

Is the purpose of econometrics not too ambitious? Indeed it is not always so easy to link mathematics, economics and statistical data.

As a better link to the real world *Econometric Models* were developed, being a mathematical formulation of a simplified but consistent global picture of economic reality, which is assisted by a group of observed phenomena. In 1936 J. Tinbergen elaborated the first econometric model namely for the Netherlands[36]. Afterwards in 1939 he published an econometric model for the United States[37].

NOTES PART 1 CHAPTER 1

[1] T.R.Malthus, *An Essay on the Principle of Population* (1803), edition Everyman's Library, London-New York, vol.i, 18-19.

[2] N.W.Senior, *An Outline of the Science of Political Economy*, Encyclopaedia Metropolitana, 1836, 131-132.

[3] A. Marshall, *Principles of Economics*, 1890, 8th ed.1920, MacMillan, New York, fourth printing 1952, Book III: on Wants and their Satisfaction, 83.

[4] H.H.Gossen, *The Law of Human Relations and the Rules of Human Action derived therefrom*, translated by R.C.Blitz, MIT Press, Cambridge (Mass), 1983, 3.

[5] W.S.Jevons, The Theory of Political Economy, 1871,304.

[6] H.H.Gossen, *op.cit.*, 3.

[7] H.Guitton, *Statistique et Econométrie,* 1959, Dalloz, Paris, 9.

[8] A.Smith, *The Wealth of Nations*, (1776-78) ed. Everyman's Library, London-New York, 1950.

[9] E.Roll, *A History of Economic Thought*, Prentice-Hall, New York, 1952, 248-324.

[10] T.R.Malthus, *op.cit.*, 18-19.

[11] "One hundred and fifty years ago, when Quesnay first published his famous schema, his contemporaries and disciples acclaimed it as the greatest discovery since Newton's laws. The idea of general interdependence between the various parts of an economic system has become by now the very foundation of economic analysis". W. W. Leontief, *Quantitative Input and Output Relations in the Economic System of the United States*, The Review of Economic Statistics, vol. XVIII, August 1936, 105.

[12] A. Marshall, *Principles of Economics*, 1890, 8th ed.1920, MacMillan, New York, fourth printing 1952, 369.

[13] For Keynes criticizing Marshall see: R. F. Harrod, *The Life of John Maynard Keynes*, London 1951, 117. For Keynes as Marshallian, see: H. G. Johnson, *The General Theory after twenty-five years*, American Economic Review, May 1961, 1-25. J. R. Hicks, *Value and Capital*, 2nd. ed. Oxford, 1953, 3.

[14] Under these circumstances it is rather strange that Keynes called his main work a *General Theory*, namely: J.M.Keynes, *The General Theory of Employment, Interest and Money*, London, first edition 1936, 176.

[15] For this first period of the life of Keynes, see the book of the biographer and previous student of Keynes, Sir R. F. Harrod: *The Life of John Maynard Keynes*, London 1951, 1-176.

[16] J. M. Keynes, *The General Theory of Employment, Interest and Money*, chapter 14: "the Classical Theory of the Rate of Interest" London, first edition 1936, 176.

[17] R. F. Harrod in a letter to W.K.Brauers (the author) on October 27, 1961, due to the fact that the author had remarked that Sir Harrod never mentioned the name of Walras in his famous biography on Keynes. In the same letter Sir Harrod added:" I have no doubt that Keynes knew the name of Walras. He certainly looked at Walras when writing the General Theory.................. He could certainly read French quite well. He read a good deal of continental work, including German, at that time. To illustrate, I recall a gathering in his room, when we were discussing the ideas of Mises. After a certain amount of equivocation he finally admitted that he himself had never read Mises, but one of those present said 'Oh yes you have' and jumped up and produced a review of Mises by Keynes in the Economic Journal before 1914, which everyone agreed was a good and fair review".

[18] P. A. Samuelson, *Foundations of Economic Analysis*, Cambridge (U.S.A.)1953, 314.

[19] H. Guitton, *Statistique et Econométrie*, op.cit., 475-478.

[20] P. A. Samuelson, *op. cit.* 311.

[21] J. M.Clark, *Economic Essays contributed in honor of J. B . Clark*, New York, 1927, 46-70.

[22] J. R. Hicks, *Value and Capital*, Oxford,2nd edn.1946, 115.

[23] Hicks writes that in mechanics dynamics means power corresponding to the meaning in old Greek. Also in politics, dynamic would mean powerful action. In common parlance dynamic could mean movement such as modern art being dynamic, *op. cit.*, 115.

[24] E. Roll, *A History of Economic Thought*, New York 1952, 471.

[25] J. Tinbergen, *Economic Policy: Principles and Design*, Amsterdam 1956,4-5.

[26] "The microtheory runs in terms of firms and the macrotheory in terms of industries". H.Theil, *Linear Aggregation in Input-Output Analysis*, Econometrica, January 1957,111.

[27] "There are, on the one hand, micro models, in which a large number of 'compartments' is assumed to exist (e.g. firms and households) and where, for example, a distinction is drawn between a large number of different goods. On the other hand there are macro models where such distinctions are not made, but certain features common to all compartments are considered, perhaps in more detail than would otherwise have been possible". *Economic Policy: Principles and Design*, Amsterdam 1956, 28. Afterwards the author develops what he means by 'common to all compartments', for instance: national income, national expenditures, the general price level, public and private expenditures, taxes, money, balance of payments, full employment, the general wage level, exchange rates etc. On the contrary the detailed description of public finance belongs to micro-economics (*op. cit.*, 40-47).

[28] For instance Balderston and Whitin state that aggregation makes the difference between micro- and macro-economics. It is a question of degree rather than of kind" J. B. Balderston and T. M. Whitin, *Aggregation in the input-output model*, Economic Activity Analysis (ed. O.Morgenstern) New York 1954, 80.

[29] R. F. Harrod, *op .cit.*, 502.

[30] It is worthwhile to mention some other important work in mathematical economics not having the pretension to be complete. P. A .Samuelson, *Foundations of Economical Analysis*, Cambridge (U.S.) 1947. R. G. D. Allen, *Mathematical Economics*, London 1957. W. J. Baumol, *Economic Theory and Operations Analysis*, Englewood Cliffs N.J. 1961. R. G. D. Allen, *Macro- Economic Theory*, London 1967. A. C .Chiang, *Fundamental Methods of Mathematical Economics*, Tokyo 2nd edn.1974. Confusion is possible with what is called: 'mathematics for economists' as in: R. G. D. Allen, *Mathematical Analysis for Economists*, London 1956.

[31] J. Tinbergen, *Econometrie*, 2nd ed. Gorinchem (Nl.)1942, written in Dutch in the beginning of the thirties, was probably the first handbook on econometrics.

[32] R. Frisch, *Editorial*, Econometrica vol. I, January 1933.

[33] J. F .van Waterschoot, W. K. Brauers, A. P. Sümer et al. *A Bank of Econometric Knowledge, Feasibility Study*, University Leuven, Louvain 1974, 1.

[34] Hoel maintains that the collecting and summarizing of data is usually called descriptive statistics, whereas the source of the data is called statistical inference" (P. G. Hoel, *Elementary Statistics*, New York-London, 1971, 2. A much more restricted definition is given by W. L. Hays when he declares that "statistical inference involves statements about probability" (W. L. Hays, *Statistics for the Social Sciences*, 2nd.edn. London, 1974, 47.

[35] R. Roy, *François Divisia, 1889-1964*, Econometrica, vol.33, N°3, 1965, 637.

[36] J. Tinbergen, *Selected Papers*, Amsterdam 1959, 37-84.

[37] J. Tinbergen, *Statistical Testing of Business Cycle Theories*, League of Nations, Geneva, 1939.

CHAPTER 2

SUCCESS AND SHORTCOMINGS OF ECONOMETRIC MODELS

Walters describes the operational task of the *Econometrician* as taking the theory of economics and providing some measurements. This means that he must collect data, characterize them, for instance as being elastic or inelastic. These empirical parameters are then used to predict events. Afterwards, comparison with the actual outcome will result in a test of effectiveness[1].

2.1 SIGNIFICANCE OF ECONOMETRIC MODELS

In order to fulfill his task the econometrician has to overcome many difficulties, such as: bureaucracy, the margin of error of each statistical work, the changing attitude of individuals, external influences such as climatic conditions, war, war menace, political attitudes of foreign governments, uncertainty of the financial markets and world business fluctuations. This influence of the world financial and business fluctuations makes an econometric model, as basis for an economic year plan of a small country like the Netherlands, so vulnerable[2].

Econometric models seem to be the most successful in periods of continuous economic growth such as in the golden sixties. This was the case with the econometric model that formed the base for the 1960-1964 plan in Belgium. Afterwards this planning effort, if based on an overall econometric model, became much more doubtful[3].

For some the dictum: "small is beautiful" would also be of application for econometric models. Large models sometimes with hundreds of equations seem to be less effective than small models, based on some generalized Keynesian type variables such as income, consumption, investment and savings[4]. The difficulty in forecasting turning points is a weak point for econometric models[5].

As multiple regression projections form the core of econometric models, their observations from the past may lead to explosive or to nonsense results for the future. This may happen if the model remains unclear to a small or a large extent about dependencies. Indeed an explosive or a nonsense result could occur with the autoregressive form or the reduced form of the models.

As a reminder, in the *Autoregressive Form* only one dependent variable as a function of the lagged variables of that dependent variable appears per equation[6].

The single equation regression model under its autoregressive form seems to be the oldest type among the econometric models[7].

In the *Reduced Form* representation each endogenous variable depends on all lagged variables and on all the exogenous variables[8]. In order to avoid an explosive process the use of the recursive form is helpful. In the *Recursive Form* each endogenous variable depends on all lagged variables, all exogenous variables and the endogenous variable, all of which were solved in the previous equation. Consequently, the first equation is of the reduced form and the following equation uses the outcome of the first equation etc., i.e., a stepwise operation[9]. The problem however rises that not only one but many alternative recursive models are possible depending on the choice of the first equation. Shapiro found that out of 19 macro-econometric models from Eastern Europe over the period 1960-1975, four were of the recursive form, three partly recursive and the others of the reduced form[10].

The *Structural Form* is the most complete form and the best protected against an explosive result. At the same time the structural form provides a good synthesis of interdependence in the following way:

$$
\begin{aligned}
x_{1,t} &= S_1(x_{2,t},\ldots,x_{v,t},\mathbf{x}'_{t\text{-}\delta},\mathbf{z}'_t) + \omega_{1,t}\\
x_{2,t} &= S_2(x_{1,t},x_{3,t},\ldots,x_{v,t},\mathbf{x}'_{t\text{-}\delta},\mathbf{z}'_t) + \omega_{2,t}\\
\ldots &\quad \ldots\ldots\ldots\ldots\\
x_{v,t} &= S_v(x_{1,t},\ldots,x_{v\text{-}1,t},\mathbf{x}'_{t\text{-}\delta},\mathbf{z}'_t) + \omega_{v,t}
\end{aligned}
$$

the x's represent all the endogenous variables;
$\mathbf{x}'_{t\text{-}\delta}$ the row vector of all the variables x being determined lagged one or more time periods;
$\mathbf{z}'_t$ represents the row vector of all the exogenous variables;
the ω are random variables, which are jointly distributed.

Starting with the concept of a structural form representation means, however, the recognition of formidable estimation problems[11]. Some authors tried to come to alternative models with a more realistic content, not always with much success. Lippincott found that mathematical demonstrations like the Walrasian simultaneous equation system, of a partly structural kind, could substitute the market economy[12]. Walras rather preceded later developments of that kind by Pareto and Barone[13].

Léon Walras (1834-1910) is considered as the father of mathematical economics. Walras sets up a general interdependent system and as a mathematician he takes an undetermined number, namely n, of factors of production[14]. This number could be one (labor Ricardo-Marx) or three (labor, nature, capital A. Smith), four (the entrepreneur as a fourth factor) or more.

The undetermined number of products m is also considered and all factors of production and all products have a price.

First, a system of n equations of total supply of the factors of production is given, each equation being a function of all the prices of the factors of production and of all the prices of the products.

Second, a system of m total demand equations of products shows that each demand is a function, once again, of all the prices.

The respective quantities of each of the factors of production which enter in the production of a single unit of each product are called: *Fabrication Coefficients.* This leads to another system of n equations, where demand is equal to supply. Indeed, each supply of a production factor equals all demands multiplied with the corresponding fabrication coefficients.

Another set of m equations expresses that the price of each product equals its cost price, viz., as a function of the factors of production with their prices. At this occasion one of the final products is taken as a basis. Therefore the price of one of the products is known and put equal to one.

One of the equations in the second set of equations is redundant, due to the fact that total demand for all products has to be equal to total supply. This means that one equation can be dropped out. In total it signifies that $2m + 2n - 1$ equations are present in order to determine $2m + 2n - 1$ unknowns:

1) the n quantities of total supply of the factors of production;
2) the n prices of the factors of production;
3) the m quantities of total demands of the products;
4) the $m - 1$ prices of these products.

Walras assumes that the fabrication coefficients are a priori fixed. This is done "ceteris paribus", which means in order to simplify the problem. He is aware of the fact that the fabrication coefficients are in reality not constant. Therefore in a second part he considers them as being variable e.g., geographically between developed and developing countries or in time.

In this second stage the respective quantities of each factor of production, which enter in the production of one unit of each product, are determined at the same time as the prices of the factors of production, under the condition that the cost price of the products has to be a minimum. Walras expresses this condition by a system with as many equations as there are fabrication coefficients to determine.

Walras used his system as an illustration of his theory of interdependence. He also showed in this way how mathematics can be used in economics. Later on the question was posited if the Walrasian system could replace the market mechanism, for instance, in a collectivistic economy. It is true that the capacity of modern computers will make enormous calculations possible. The statistical and econometric data for the equations, however, do not yet exist. Moreover the analysis is static: neither investments, nor interest rates are considered as an instrument of choice between the present and the future. This is also one of the main difficulties in a collectivistic system in which the market economy is absent. Not too many chances are given to a replacement of the market mechanism.

The *Input-Output* system (I/O) is another interdependent system, which is much more operational and statistically rich, but this system does not fulfill all conditions of a Walrasian system. Indeed, in order to be complete the input-output system has to evolve from an open to a closed system. Moreover two kinds of input-output systems could exist: one where the quantities are unknown, another where the prices are the unknowns, but they are not operating together.

Anyway, the Input-Output models bring a practical solution to the Walrasian system. They are in fact a set of interrelated production functions.

2.2 DEFINITION OF I/O AND ITS RELATION TO THE WALRASIAN SYSTEM

The birth of Input-Output is not so recent as generally accepted. The first article of Leontief on his method dates back to 1936[15]. It did not draw much attention. However, his book: "The Structure of American Economy (1919-1929): an empirical Application of Equilibrium Analysis" published in wartime (1941), was out of print in a short time. Was it mere chance or due to the fact that the war economy makes government interference of importance? In 1951 appeared a second, enlarged edition of this book. Professor Leontief depicts his method as:

> "an attempt to apply the economic theory of general equilibrium - or better, general interdependence - to an empirical study of interrelations among the different parts of a national economy as revealed through co-variations of prices, outputs, investments, and incomes"[16].

The system of Leontief is a simplification of Walras' general equilibrium system, but this time tested in reality. The criticisms of von Mises, von Hayek and Robbins and the defense of Barone and Taylor in the controversy concerning the utility and solution of Walras' system, is centralized around the market mechanism[17], whereas, Leontief has scrutinized the utility of the system from the point of view of production.

The backbone of Leontief's system is a matrix for the whole economy, wherein the outputs are rows and the inputs are columns. Both inputs and outputs, expressed in values or in quantities, are interrelated through a set of technical coefficients, of much importance in the system. In practice, however, all Leontief matrices have been expressed in money-terms and not in physical units.

It is quite possible that one has the impression from this study that by open economy is meant an economy which is influenced by a considerable amount of international trade. Leontief's system, although not originally so, became an open system in the sense that consumption, either household consumption or government expenditure, is not explained inside the system. Also, international trade, stocks, investments and economies of scale is kept out of the system.

Leontief's matrix, which was extended to 450 production sectors at a certain moment, gives a powerful statistical picture for an economy, with a large advantage over the sometimes confused statistical yearbooks, published by many countries. The input-output system is mainly a study of inter-economic relations. As such it can reinforce the national accounts. Indeed national accounts are consolidated value-added and income statements, in which all double-counted items are dropped out, so that a fair picture is obtained of what is going on in the economy. The input-output study helps to cover a gap in the national accounting system by giving the production flows. In the same way a study of *Moneyflows* will show all the moneyflows in the economy[18]. Professor R. Stone, later a Nobel Prize Winner, has given the mathematical formulation of all these interrelationships[19]. He considers the available data of a national economy in a *Social Accounting Matrix* (SAM).

2.3 SAM AND GENERAL EQUILIBRIUM MODELS

Richard Stone was much impressed by the production flows as represented in the Input-Output matrix[20]. Subsequently, he cast the national accounts of the United Kingdom also in matrix form. This representation became an important element in the United Nations revised system of national accounts. Finally, he combined both matrices in a Social Accounting Matrix[21].

Moneyflows data have to be integrated in SAM[22]. International organizations, such as the International Labor Office and the World Bank promote the application of SAM, mostly for developing countries[23]. SAM is also seen as a basis for modeling[24].

Separate from this work, first studies on *General Equilibrium Models* (GEM) were made as early as 1960, e.g., by Johansen[25]. Similar publications till 1982 on GEM are considered by some as too expensive, difficult, of limited applicability and weak flexibility. Therefore they prefer the name of *TV (Transaction Value)*. SAM remains at the basis of TV[26]. More recently the name changed to *Computable General Equilibrium Models (CGE models)*, as the approach is facilitated by the advances in computerization[27].

The CGE models go much farther than the interrelated production functions of input-output. They also cover consumption, income, savings, investments and the moneyflows. For forecasting and economic policy the CGE models encounter the difficulties, which were mentioned earlier, related to the reduced, recursive or structural forms of the simultaneous equation systems.

2.4 THE STABILITY OF THE TECHNICAL COEFFICIENTS AND THE INVERSE MATRIX

More or less in comparison with the fabrication coefficients of Walras, Leontief introduced T*echnical Coefficients*. These technical coefficients are, however, a result of aggregation. Indeed, these technical coefficients are based on sectors and not on products as in the fabrication coefficients. A technical coefficient shows the ratio between the input of a sector and the production of another sector. For instance, the technical coefficient of petroleum products for the road transport sector shows the input of petroleum products for the road transport sector divided by the production of the road transport sector. Due to the aggregation and adding problems these aggregates have to be in money terms and not in physical units.

And the stability of these technical coefficients in comparison with the stability of the fabrication coefficients of Walras?

It would be interesting to look for any regularity over time in the behavior of the technical coefficients. Unhappily, statistical information on the technical coefficients is too limited to set up time series for a sufficiently long time. Leontief made the following important statement:

> "Anybody familiar with the economies of industry and agriculture knows that most of the major technological developments are well discernible far in advance of their actual culmination, while minor innovations are frequently planned by leading enterprises well ahead of their general adoption"[28].

His experiences for the 1919-29 period and his *Backward Prediction* from 1939 tot 1929[29] are very well known. Cameron cited stability for the technical coefficients for five to ten years[30]. We do not think that such a long period of stability was acceptable for Western Europe immediately after the Second World War. Indeed, Western Europe had to catch up with the industrial development of the United States. We found out that the first Belgian input-output table (1953) was out of date in 1958[31]. Therefore, we set up a 1958 national table[32]. The fact, that the European Union (Eurostat) asks the member countries to elaborate an input-output table for each five-years since 1959, shows the necessity for this updating for all the member countries. On the other side the period 1973-1975, with the first energy crisis, may be considered as a trend breaking period.

Contrary to the opinion of Leontief, we think that nowadays the stability of the technical coefficients is rather a result of a large aggregation of many production processes. Indeed, stability of coefficients means linearity and constant returns to scale. Production processes, however, are characterized by constant returns to scale, but also by increasing or decreasing returns to scale. Moreover, over time production functions may change. By aggregation, opposite evolutions may compensate each other and lead to the mentioned stability, which is the empirical finding over so many years. In other words, input–output forecasting is applicable for macro-, but not for microeconomics[33].

Once the stability of technical coefficients is accepted, it is not difficult to explain the inverse models as expressed in matrix algebra.

Suppose:

$\boldsymbol{X}$ as the matrix of total and sectoral intermediate demand

$\boldsymbol{x}$ as the vector of total deliveries originating from internal production or from imports

$\boldsymbol{I}$ as the identity matrix

$\boldsymbol{a}$ as the matrix of technical coefficients defined as:

$$\boldsymbol{a} = \boldsymbol{X}.\hat{\boldsymbol{x}}^{-1}$$

$\boldsymbol{y}$ as the vector of sectoral final demand,

$(\boldsymbol{I}-\boldsymbol{a})$ as the technological matrix;

then:

$$(\boldsymbol{I} - \boldsymbol{a})\,\boldsymbol{x} = \boldsymbol{y}$$

and if: $|\boldsymbol{I}-\boldsymbol{a}| \neq 0$

then:

$$\boldsymbol{x} = (\boldsymbol{I} - \boldsymbol{a})^{-1}\boldsymbol{y}$$

The inverse model, based on this inverse matrix, is very often used for prediction and planning purposes starting from a possible or desired final demand.

Very important for these purposes is the use of the *Activity Multiplier*, which is defined as follows:

$$M_i = \sum_{k=1}^{k=n} A_{ik}$$

with: M_i as the *Activity Multiplier*. For one unit of final demand i all the n sectors of the economy are activated.

A_{ik} is the inverse coefficient from the inverse matrix, whereby sector i activates sector k.

If the direct and indirect influence of the imports is omitted, the *Internal Activity Multiplier* is obtained. The internal activity multiplier effectively measures the influence of the increase of final demand on the national economy[34].

2.5 STRUCTURAL CHANGES AND INPUT-OUTPUT

Since the first energy crisis of 1973-1975, the energy inputs in the input-output sectors were disturbed and essentially changed. On the other hand, for the first time since the Second World War, very significant negative growth rates were observed in the United States, Japan and Great Britain in 1974 and in the United States, Italy and Belgium in 1975, while Western Germany had a zero-growth rate in 1975. It is quite clear that technical coefficients are seriously altered since 1973-1975. For that reason the comparison of input-output figures from before 1975 with actual figures is doubtful. The consideration of trend breaks is also very important for developing countries. Tremendously high inflation and large changes in interest rates in the countries of the industrialized world, are other important events for making the Input-Output tables invalid. As the tables are in value terms, it is at least necessary to switch over to constant prices.

Methods like RAS try to counter these problems[35]. RAS however is less involved in structural changes. Therefore we developed PES[36]. In the PES method three stages are foreseen: Price variations, Extrapolations (quantities) and Structural Changes.

1. Price Variations

Starting with the intermediate matrix of an existing input-output table $({}_1X)$, a new matrix is derived :

$${}_pX = {}_1X \, . \, {}_tP$$

with: ${}_tP$ as the matrix of ratios between prices in period t and prices in period 1 for each element of the given matrix.

2. Quantitative Extrapolations

$${}_qX = {}_pX \, . \, {}_t\hat{Q}$$

with: ${}_t\hat{Q}$ as the diagonal matrix of the ratios between the totals of intermediate demand per sector in period t and period 1.

Totals per sector of intermediate demand for period t are, however, not available, but final demand, imports, value added and national production per sector are available (marginal data). How is intermediate demand per sector for period t calculated?

a) Row totals of intermediate demand per sector for period *t*

$$r = x + m_{ik} - y$$

$$r = x + (m_{ik} + m_{i,n+1}) - (y + m_{i,n+1})$$

with: r = the vector of the row totals of intermediate demand per sector

x = the vector of national productions (available)

y = the vector of final demand of internal origin

m_{ik} = the vector of imports of intermediate demand

$m_{i,n+1}$ = the vector of imports of final demand

$(m_{ik} + m_{i,n+1})$ = the vector of total imports per sector (available)

$(y + m_{i,n+1})$ = the vector of total demand (available)

b) Totals of columns

$$k = x - v$$

with: k = the vector of column totals of intermediate demand per sector

x = the vector of national productions (available)

v = the vector of values added (at market prices or at factor costs depending of the position which is taken concerning indirect taxes) (available).

Starting with the intermediate matrix of the existing input-output table $({}_1X)$, prices and marginal data of period t, an iterative procedure is followed in order to reach a correspondence between rows and columns[37]. Substitution effects (in the rows) and technological effects (in the columns) are automatically introduced in the new matrix as compared to the situation which is shown in the matrix of period *1*.

3. Structural Changes

If important changes occurred in the most important sectors of the economy of the country, then these sectors are studied separately in monographs. The results will replace columns and rows of that sector in ${}_qX$.

The application of PES means cost and time saving[38].

2.6 BENCHMARKING AND INPUT-OUTPUT

For a developing country the stage of structural change is extremely important. Indeed an introduction of a new industrial firm may completely and suddenly upset the technical coefficient. This could be the case for the installation of a new car assembly plant. This new plant will completely change the technical coefficients of the production sector of the means of transportation or will even create a new sector. At that moment the developing country can do benchmarking, i.e., using the experience of similar situations in other countries. Also developed countries can learn from others.

What about the use of engineering data? At that moment a translation by sector specialists is needed. Indeed engineering data are purely technical and/or belong to microeconomics. Unhappily sector or branch specialists are only present in very large countries, like in the United States.

2.7 MULTIPLE OBJECTIVES AND MULTIPLE UNITS IN INPUT-OUTPUT ANALYSIS

Like the other econometric models, input-output is based on facts and tendencies of the past. However, input-output concerns rigid macro production functions, which do not change overnight. If these production functions are ascertained in the past, they are valuable for a certain time, for instance five years, unless sudden structural changes occur as was mentioned above.

Leontief, the pioneer of input-output, wished to use different physical units, but the need to inverse the matrices made for him the translation in money-terms necessary[39].

Two examples will be given in order to show that physical units can be introduced. In addition, multiple objectives are considered, but based on interdependence.

Introduction of Employment

In his doctoral thesis, the author set up a model for international integration to maximize final demand and to maximize labor productivity[40]. Final demand would have to reach at least a certain goal and labor coefficients have to be minimized under the constraint, however, of full employment. A *Labor Coefficient* was defined as the ratio of the quantity of labor in a sector, expressed in person hours, and the production in money terms of that sector.

Introduction of Pollution Abatement

The model, which was developed for the Economic Commission of Europe of the United Nations, described pollution and pollution abatement in physical terms beside a traditional input-output framework in money terms[41]. On the one side, there was the maximization of final demand and on the other side the minimization of pollution. Introduction of pollution and pollution abatement is not an easy job, as many factors have to be taken into consideration.

The following types of pollution were taken into consideration, expressed in tons per year:

in industry:

- the pollution generated by the combustion of refined petroleum products (SO_2);
- the pollution generated by the combustion of coal (SO_2);
- solid waste;

in final demand:

- water pollution (Biochemical Oxygen Demand and suspended solids);
- air pollution by motor vehicles (CO, NO_x, SO_2, particulates, hydrocarbons);
- air pollution by combustion of fossil fuels (CO, NO_x, SO_2, particulates, hydrocarbons);

Three pollution abatement processes were studied at that time: desulfurization of petroleum products, desulfurization of flue gas from coal and solid waste treatment, all expressed in tons. A pollution abatement process may create its own pollution, a difficulty that has to be taken into consideration.

After the *Polluter pays Principle*, the cost prices of abatement are attributed to the polluting sector, which have to increase their prices[42]. This shadow price system is taken into account by the PES method (see 2.5 of this Chapter). Furthermore, complete pollution abatement is difficult to attain. Indeed, perfection is difficult to achieve unless one is not afraid of a prohibitive cost. Some risks have to be taken. In fact, there are three dimensions: degree of pollution abatement, costs and uncertainty.

Each pollution abatement process is taken as a sector in the input-output terminology. As pollution and pollution abatement are defined in tons and the other sectors of input-output in money terms, technical coefficients cannot be calculated. However, the obtained square matrix can be inversed and the direct and indirect effects calculated. By using *Labor Coefficients*, total repercussions on employment can be calculated.

The applications on employment and on pollution abatement show that consideration of multiple objectives is possible if interdependent relations exist between them.

2.8 CONCLUSIONS ABOUT TRADITIONAL MODEL BUILDING FOR APPLICATION

The publication of input-output tables is most of the time too late in order to be operational for forecasting or planning. Is this the reason why the international conferences on input-output, which are held once every two years, and "Economic Systems Research", the journal of the International Input-Output Association, have few contributions on forecasting or planning[43]? It is true that in this domain input-output was much used in the previously socialist countries of Eastern Europe. These planning efforts were, however, not very successful, as is generally known.

It was remarked that SAM includes input-output and national income and product accounts in a consistent framework. The estimation is a difficult problem. Indeed Input-output tables are usually prepared only every five years, whereas national income and product data are produced annually. Additionally these estimations are produced with a lag[44].

In fact most of the SAMs drop input-output and only rely on national accounts and on other external data.

Structural changes remain a hot issue for SAMs, for General Equilibrium Models and for econometric models. Nerlove is right when he observes:

> "In order to estimate and apply an aggregative econometric model, a period must be found over which it can plausibly be assumed that no structural changes have occurred or that those which have occurred are of an exceedingly simple variety"[45].

He continues by saying:

> "Quarterly models are analytically more useful than annual models".

On the other hand there are very long trends, which have to be taken into account. It mainly concerns population trends: population growth, size of the family, pyramid of population structure, growth of the labor force, learning and education, legislation on pensions, etc. Therefore it is very strange that developing countries, like Tanzania, Swaziland and others, draft development plans for twenty-five years without mentioning these population studies.

All kind of external data may influence the outcome of the econometric models such as climatological conditions and epidemics, terrorism and wars, elections and other political changes, and changes in world trade patterns. In the last case small countries with large international trade will much depend on decisions in other countries. For instance, Belgium exports 37% of its national production[46]. In other words, *Uncertainty* is a weak point for econometric models.

There is too much extrapolation from the past in SAM, GEM and in econometric models. They have to be incorporated in a broader framework for prediction or policy purposes.

Why did we devote so much time to *Interdependence Systems*, such as the Walrasian system, Input-Output Analysis, SAM and General Equilibrium Models? The importance will become clear when we like to diminish the number of independent objectives later on. Indeed, a large number of independent objectives is not very operational for a global optimization.

Also *Dependence Relations* help to limit the number of independent objectives. A combustion furnace for waste products causes cancer for the people living in the neighborhood. Cause-effect relations are, however, most of the time not singular. The cancer, in the example, can also be caused by traffic on the nearby highway. Also cancer can be hereditary and is only declared most of the time after many years.

Statistical interdependence, such as the outcome of *Correlation* and *Regression*, is not always very useful to limit the number of objectives. Indeed these relations are a result of observations in the past (*Time Series*), which limits their applicability in a world of uncertainty. They could be combined with *Cross-Section Data*, which shows a picture of today, for instance through sampling.

Anyway, in the optimization process with multiple objectives interdependence or dependence of the objectives can play an own role[47].

Domains exist where dependency or interdependency are present but difficult to formulate. We shall call it *Quasi-Interdependency*. Quasi-Interdependency will be the subject of the next part of this book under the form of nominal methods, cross-impact, simulation, sensitivity analysis and scenario writing.

NOTES PART ONE CHAPTER 2

[1] A.A.Walters, *An Introduction to Econometrics*, London 1968, p. 21

[2] Initially, the Central Planning Bureau of the Netherlands was one of the pioneers of economic yearly plans; see therefore: *First Memorandum on the Central Economic Plan 1946 and National Budget* 1947, CPB, The Hague, September 1946.
Scope and Methods of the Central Planning Bureau, CPB, The Hague, August 1956.
The *Centraal Economisch Plan 1955*, March 1955 gives more information on technical aspects of short-term planning.

[3] See e.g.: R.Maldague, *Planification et stratégie économique*, Tijdschrift voor economie en management, University Leuven 3-4,1985, 415-430.
Nowadays the Belgian Planning Office uses econometric models only as a starting point, but the results are corrected with outside information, such as through *Nowcasting* for more recent data. In addition, the judgment is asked of experts before a final result is published. This way of acting is a copy of what already happened in the OECD, the IMF and the World Bank.

[4] We found a first trace of this dilemma in the debate of 1954 between H. Barnett, A. W. Marshall and S.Lebergott (Long- Range Economic Projection, Studies in Income and Wealth, A Report of N.B.E.R., Princeton University Press, Princeton 1954, 191-232). Stekler compared six quarterly models with two naive models suggesting that the quarterly econometric models were not so successful in forecasting economic activity. H.O.Stekler, *Forecasting with econometric models: an evaluation*, Econometrica, 3-4,1968, 437-463.

[5] Theil's inequality coefficient could be helpful here: H.Theil, *Economic Forecasts and Policy*, North Holland, Amsterdam 1958, 31-35.

[6] P.Kennedy, *A Guide to Econometrics*, Blackwell,Oxford, 1998, 44 and 155.

[7] G.H.Orcutt, *Microeconomic Analysis for Prediction*, in: Forecasting on a Scientific Basis, Instituto Gulbenkian de Ciência, Lisbon 1967, 71.

[8] P.Kennedy, *op. cit.*, 158.

[9] A.C.Darnell, *A Dictionary of Econometrics*, Edward Elgar, Cheltenham, 1995, 343-346.
P.Kennedy, *op. cit.*, 169.

[10] H.T.Shapiro, *Macroeconometric Models of the Soviet Union and Eastern European Economies. A Tabular Survey*, Econometrica, vol.45, 1977, 1747-1766.

[11] P.Kennedy, *op. cit.*, 157-158.

[12] B.E.Lippincott, O.Lange and F.M.Taylor, *On the Economic Theory of Socialism*, The University of Minnesota Press, Minneapolis, 1948, 12-13.

[13] V.Pareto, *Cours d'économie politique*, Lausanne 1897, vol.II,364 and ff.
O.Lange in: B.E.Lippincott, O.Lange and F.M.Taylor, *op.cit.*, 59.

[14] L.Walras, *Eléments d'économie politique*, (1874) Paris 1952, p.208 a.f.

[15] W.W.Leontief, *Quantitative Input and Output Relations in the Economic System of the United States*, The Review of Economics, XVIII, August 1936, 105-125.

[16] W.W.Leontief, *The Structure of American Economy, 1919-1939. An Empirical Application of Equilibrium Analysis,*. Second enlarged edition, Oxford University Press, New York, 1951, 3.

[17] B.E.Lippincott, O.Lange and F.M.Taylor, *op. cit.*
F.A.von Hayek, *Collectivist Economic Planning*, Routledge, London, 1935.

[18] For the study of moneyflows, see the pioneering work of M.A.Copeland: *A Study of Moneyflows in the United States*, National Bureau of Economic Research, New York,1952.

[19] R.Stone, *Input-Output and the Social Accounts*, International Seminar on Input-Output Analysis, Varenna, Italy, 1954, mimeo.
R.Stone and J.E.G.Utting, "The Relationship between Input-Output Analysis and National Accounting". *Input-Output Relations*, Proceedings of a Conference on inter-industrial relations held at Driebergen, Holland, Leiden 1953, 195.

[20] R.Stone and G.Croft-Murray, *Social Accounting and Economic Models*, Bowes and Bowes, London, 1959,25-26.

[21] R.Stone and G.Croft-Murray, *op. cit.* 29.

[22] G. Pyatt, *The Sam Approach in Retrospect and Prospect,* Warwick Economic Research Papers, N°290, Department of Economics, University of Warwick, 1987, 18.

[23] "By now SAMs have been designed and quantified for well over 30 countries". G.Pyatt, *op. cit.*,18.

[24] G.Pyatt, *op. cit.* ,32-39.

[25] L.Johansen, *A Multi-Sectoral Study of Economic Growth*, North Holland, Amsterdam, 1960.

[26] B.Decaluwe, W.Grais, *La méthode T.V. pour l'élaboration de modèles d'équilibre général*, Research Department, World Bank, 1983.

[27] D.Vongpradhip, A CGE Model with Real and Financial Sector Linkages, Thailand Development Research Institute Foundation, Bangkok, 1987. This study is based on SAM and makes a strong link with financial data. Others refer to the link between SAM, CGEM and TV: La méthode de construction des modèles calculables d'équilibre général de type TV, Centre de recherche et développement en économique, Université de Montréal, 1984.

[28] W.W.Leontief, *The Structure of American Economy, 1919-1939*, op.cit., 1951, 152.

[29] W.W.Leontief, op.cit., 153

[30] B.Cameron, *"The Production Function in Leontief models"*, Review of Economic Studies, 1953, XX, N° 1, 62-69.

[31] W.K.Brauers, *"De evolutie van de input-output technische coëfficiënten met bijzondere toepassing voor België"*, Revue belge de statistique et de recherche opérationnelle, May 1965,11-40.

[32] W.K.Brauers, *"De opstelling van een input-output tabel voor de Belgische economie in 1958"*, Cahiers Economiques de Bruxelles, N° 21, 1964, 115-136.

[33] We define *Macroeconomics* in terms of economic sectors eventually aggregated to a national economy, international or world economy; *Microeconomics* in terms of firms. Therefore we follow Theil: "The microtheorey runs in terms of firms and the macroeconomy in terms of industries". H.Theil, *Linear Aggregation in Input-Output Analysis*, Econometrica, January 1957, 111.

[34] How to measure the internal activity multiplier is explained in: W.K.Brauers, *Prévisions économiques à l'aide de la méthode Entrées-Sorties*, Economica, Paris, 1995, 48-54.

[35] *Input-Output Relationships 1954-1966*, A Programme for Growth, N° 3, Department of Applied Economics, University of Cambridge, September 1963, 27-41.

[36] PES was used for several applications: 1) the first Belgian input-output tables were made by a research group of the University of Brussels for 1953. A PES extrapolation was made for 1958 (see footnotes 25 and 26 above); 2) for 1964, a PES interpolation, linked to provincial structural changes, was made for the province of Belgian Limburg. At the same time, the direct and indirect repercussions of the closing down of the last Belgian coal mines were estimated (*Omschakelingsplan Limburgse Mijnstreek*, LER, Hasselt (B), 1966); 3) On basis of the official I/O table for 1970 (only published in April 1975, the official table for 1975 was only published in 1983) and structural changes, a PES extrapolation was made for 1976. In this way the crisis years 1973-75 were avoided (W.K.Brauers, J. van Waterschoot, P.Van Elewyck, *De sectoriële en regionale analyse van de Belgische economie*, Center for Economic Studies, Leuven University, 1979) This 1976 table was used as a basis for interregional tables (W.K.Brauers, *The Belgian Experience in Interregional Input-Output Tables*, Center for Economic Studies, Leuven University, 1980.

It was assumed that normally in Western Europe an I/O table is valuable for five years (see Part I, 2.4 in the main text). Consequently, another five years for an updated PES table is acceptable. It can be remarked that the publication of the official Belgian tables is always too late. This remark can be made for all the countries of the European Union. It is even the case worldwide.

[37] This procedure is explained in: M.Bacharach, *Estimating nonnegative matrices from marginal data*, International Economic Review, N° 3, September 1965, 294-310.

[38] Is the step of using PES worthwhile? An illustration is given through a historical anecdote. During a meeting of the "Société Belge de Statistique", on October 29, 1964, Professor E.S. Kirschen estimated that the first Belgian input-output table for 1953, which was made by its team at the University of Brussels, needed five man-years. The representative of the Belgian Institute of Statistics estimated that the time to finish the first official Belgian input-output table for 1959 was much longer. In the mean time the author, for the first time using PES on the basis of the 1953 table, needed one man-year to finish the table for 1958. Consequently PES is not only a question of costs but also of time saving.

[39] W.Leontief, *Studies in the Structure of the American Economy,* Oxford University Press, New York, 1953, 8.

[40] W.K.Brauers, *Input-Output Analyse en Internationale Integratie*, with a preface of W.Leontief, Standaard Wetenschappelijke Uitgeverij, Antwerp-Utrecht, 1968, 201-229.

[41] J.van Waterschoot, W.K.Brauers, P.van Elewyck, *The linking of Polab with an Econometric Model*, the Economic Commission for Europe of the United Nations, Geneva, 1976.

[42] The in most countries legal *Polluter pays Principle* is maintained, although it is against the general principles of the economics of well-being. Indeed, in the first stage the polluter pays the costs of pollution abatement. In the second stage, these costs are shifted to the customers of intermediate demand, if the elasticity of their demand is low; they in their turn will finally shift the price increases to the consumers if the elasticity of their demand is low. Demand is more inelastic for low income classes than for high income classes, with the result that price increases, due to polluter abatement, are in a large part shifted to lower income classes.

Pollution abatement, however, is a social benefit as the quality of life is improved (less disease, lower mortality rate, fewer industrial accidents etc.). Hence, a tax levy for pollution abatement is logical. How much has to be paid? The tax rate will be positive correlated to the pollution abatement rate. It would depend on the consumer surplus of well-being to be realized by pollution abatement. We know that complete pollution abatement is impossible due to the asymptotic growth of abatement costs towards the ceiling of complete abatement The stakeholders will decide upon an abatement level to be maintained as a function of the social benefits produced.

[43] This journal is published quarterly since 1989: Economic Systems Research, vol.1, N°1.

[44] S.Robinson, A.Cattaneo, M.El Said, *Updating and Estimating a Social Accounting Matrix Using Cross Entropy Methods*, Economic Systems Research, International Input-Output Association, March 2001, 47-64.

[45] M.Nerlove, *A Quarterly Econometric Model for the United Kingdom*, The American Economic Review, March 1962, 155.

[46] W.K.Brauers, J. van Waterschoot, P. Van Elewyck., *op. cit.* 1979, 4.

[47] A very clear distinction between multiple objectives with or without interdependencies is made by: A.Mccowan, S.Mohamed, *A classification of decision support systems (DSSs) for the analysis and evaluation of concession project investments (CPIs)*, Journal of Financial Management of Property and Construction, vol.7, N°2, August 2002,129-134.

PART TWO

THE ADVENT OF NEW ECONOMICS

CHAPTER 1

CREATIVE AND PROSPECTIVE THINKING

Futurology is the technique of using imagination about the future. It promotes *Creative Thinking*. Futurology has not to be confused with wishful thinking, in which one hopes that his own wishes will come true. The problem of futurology is that it does not take enough into consideration the actual situation. At that moment the division line is difficult to draw where wishful thinking begins.

Prospective Thinking combines creative thinking with a sensibility for reality. In this way it could also consider extrapolations.

1.1 SOME PRELIMINARY DEFINITIONS

In order to clarify our thinking, a classification of events is necessary. The *Statistical Universe* represents the set or a subset (for instance, only concerning income) of all events. The *Statistical Population* forms a partition of the universe, namely, the set or a subset of all passed events. It is also useful to define samples and polls. A *Sample* can be drawn from a statistical population. For instance, from a population of four million households a random sample of 4,000 households is drawn. These households will keep record of consumption during one year. This subset of the population is considered significant for the whole population. A *Poll* represents a possible and significant estimation of a future subset of the statistical universe to which it belongs. Therefore the notion of *Agent* has to be introduced: an agent is an actor in the part of the population or universe under consideration. For instance, the preference of a certain number of American voters concerning a next president is considered as valuable for all American voters. It is clear that a significant number of American voters is concerned and not e.g., British voters. Consequently, an agent has to be an active member of the subset. For that reason the following interviews were wrong.

In 1976, the "European Coordination Centre for Research and Documentation in Social Sciences" in Vienna interviewed 9,000 persons in ten countries, aged between 15 and 40 at the time of interview, an age group which on the average would live at least until the year 2000[1]. More than 50 % of those interviewed were not expecting a world war before the year 2000. If the stratified sampling is correct (9000 persons, 10 countries chosen from all over the world), the majority of the world's population aged between 15 and 40 did not expect a world war before the year 2000. Besides the cost of the operation during three years (1967-1969) a large minority of the age group in question expected a world war, while the question remained unanswered whether there would be a world war or not. The assumption of the steering group that the future state of the world would largely depend on what people want it to be was purely wishful thinking. In this way we did not even know what the probability was of a possible world war before the year 2000. On such a question, there was no convergence of the opinions of the 9000 persons. Finally, are these persons aged from 15 to 40 able to make a valid judgment on such a question and is this judgment valid for the world's population at large? These people are not really "agents", but they would passively undergo a world war. In fact one had to question the decision makers in this field themselves, which is here impossible.

There is, however, something like expert knowledge, or better especially knowledgeable people, which could replace the decision makers. In the case of the world war a panel of selected politicians, generals, representatives of peace foundations and specialized think tanks and professors in international relations has to be called together.

In the following pages some methods will be explained to assist the experts. Two different kinds of methods are proposed: on the one side the open discussion methods and the nominal methods on the other.

1.2 OPEN DISCUSSION METHODS

In open discussion methods a face-to-face dialogue is essential. This is the case with committees and brain storming.

1. Committees or Round-Table Discussions

Committees or round-table discussions are not very suitable for the discussion of very broad issues. E. S. Quade outlined their main disadvantages in the following way[2]:

> "In broad problems the range of expertise required is not likely to be provided by a single individual. Almost inevitably a variety of expert advisors needs to be consulted. Experiments have shown that their best use is not the traditional one of having the issues presented to them and debated in open round-table discussion until a consensus emerges or until they arrive at an agreed-upon group position. Committees, for example, often fail to make their assumptions and reasoning explicit. Sometimes the opinions of dissenters are not even recorded. What is needed is a way to avoid the psychological drawbacks of a round-table discussion - such as the "bandwagon" effect and the unwillingness to abandon publicly expressed opinions - and thus to provide a setting in which pros and cons of an issue can be examined systematically and dispassionately".

Moreover, in committees or round table discussions the opinions of very interesting but rather shy persons are often not even heard.

2. Brainstorming

Jantsch, citing Von Fange, give the following basic rules for *brainstorming sessions*[3]:

"1. State the problem in basic terms, with only one focal point;
2. Do not find fault with, or stop to explore, any idea;
3. Reach for any kind of idea, even if its relevance may seem remote at the time;
4. Provide the support and encouragement which are so necessary to liberate participants from inhibiting attitudes".

In any case, an efficient reporting system is necessary to record the ideas presented (stenography or recording). Besides, this straight brainstorming Jantsch cites some variations[4]:

"Whereas straight brainstorming aims primarily at a harvest of new ideas, the "*buzz group*" technique seeks group consensus among approximately six people. The *operational creativity approach* introduces the refinement that only the group leader knows the exact nature of the problem and structures the discussion so as to arrive at a solution - only one is sought".

In the last variation the objectivity of the group leader is doubtful and in general brainstorming is insufficient for tackling broad problems and for obtaining judgmental data. Indeed opinions can be too divergent for a consensus to be reached.

Brainstorming must be considered too simplistic and too naive. Brainstorming is valuable for obtaining a first approximation, especially brainstorming in a meeting of experts.

Insight into alternative futures and the definition of objectives and of alternative policies for the nation, the public sector and the enterprises are provided by devices, generally computer assisted, such as delphi, nominal group techniques, cross-impact analysis and simulation. Delphi, nominal group techniques and cross-impact analysis are nominal methods, i.e., the face-to-face dialogue is absent in order to decrease subjective influences.

1.3 NOMINAL METHODS

Nominal methods, contrary to open discussion methods, are not based on face-to-face dialogues. Delphi, Nominal Group Techniques and Cross Impact are examples of nominal methods.

General Characteristics of Delphi

Delphi tries to improve either the panel or committee approach, or the questionnaire or poll approach. Delphi was so named after the Greek oracle, as it was first thought of as a tool for better forecasting. In this sense, it seems that the first experiments took place around 1948[5]. Today Delphi is no longer limited to forecasting alone. Dalkey and Helmer at RAND Corporation first used Delphi in its present form around 1953[6].

The Delphi Method is a method for obtaining and processing judgmental data. It consists of a sequenced program of interrogation (in session or by mail) interspersed with feedback of persons interested in the issue, while everything is conducted through a steering group.

The essential features of Delphi are the following:

1) the rather vague notion "persons interested in the issue", is interpreted by Quade as follows:

   "In practice, the group would consist of experts or especially knowledgeable individuals, possibly including responsible decision makers"[7];

2) the steering group treats anonymously the sources of each input;

3) inputs must as much as possible possess a single meaning and a quantitative form. The inputs with these characteristics are elicited with feedback in a series of rounds;

4) opinions about the inputs are evaluated with statistical indexes such as median and quartiles;

5) there is also a feedback of the statistical indexes with a request for re-estimation after consideration of reasons for extreme positions. The practice of Delphi reveals that after several rounds convergence is shown between the various opinions (one of the main advantages of the Delphi method);

6) there are two developments of Delphi: one is based on a meeting, the other on the sending of questionnaires. The organization of a meeting produces quicker results; the meeting, however, has to be organized in such a way that communication between the panel members is impossible. In order to increase even further the speed of the outcome of a meeting, an on-line computer could be installed. Everybody involved in the Delphi teamwork would have a desk terminal linked to a computer and would be able to look at a television screen giving the results calculated by the computer.

The disadvantage of the Delphi meeting is the local limitation of the choice of experts or especially knowledgeable individuals. Nowadays, however, teleconferences could take away this geographical limitation.

Not all applications, either in private firms or in government services, are known by the outside world. Therefore we mention known examples since the origins of Delphi.

The first application of Delphi by Dalkey and Helmer concerned the Soviet viewpoint of the estimation of the number of A-bombs necessary to take out the US industrial system[8]. Since then several hundreds of studies have applied the Delphi technique. Some of the subjects treated include: long-range forecasting (in six main areas)[9], educational planning and educational developments[10], economic forecasts and corporate forecasting[11], technological and automation developments[12], societal developments[13], employee benefits[14], regional studies[15], long-range aerospace program planning[16], the best research projects in the public interest[17]. Delphi can be used for values as well as for facts.

The Delphi experts particularly stressed the value aspect in the Delphi-judgments[18]. Value judgments were included in studies on the desirability of civil defense programs[19] and on the impact of the quality of life on public programs[20].

An Example of Application of Delphi

Let us illustrate these explanations with a detailed example. In 1970, the American "Kettering Foundation" asked "The Institute for the Future" for the best research projects in the interests of the American people with a cost per project of between ½ and 5 million $[21]. A Delphi-exercise was carried out by mail. Forty-three specialists from all branches of science participated. This group was augmented, in the first round only, in order to generate more ideas, by ten staff members of the Kettering Foundation and by ten staff members of the Institute for the Future[22].

189 proposals were produced in this way in round one.

> "These were processed as follows. Some, which evidently were of identical or closely related intent, were combined. A few were omitted, because in the experimenters' judgment they either were too obscurely worded to permit reasonably unambiguous interpretation or were not responsive to the instructions of Questionnaire 1. Finally, the majority of proposals were reformulated slightly in an effort to enhance clarity and to achieve a modicum of uniformity of presentation"[23].

165 proposals remained which formed the inputs into round two. In this round, the 43 experts gave their evaluation. The medians of the results were computed.

> "These medians were taken as numerical indicators of the panel's opinions as a group. Based on these assessments, more than half the proposals under consideration were eliminated from the final round on one of three grounds:
> - too low a probability of success;
> - too small a societal impact even if successful;
> - outside the cost-range" [24].

In the third round each panelist reviewed the median figures derived from round two and gave his final evaluation.

The ten projects that ranked highest belonged to the human sciences. The highest ranked was that on human intelligence:

> "What is needed is a new, comprehensive, up-to-date study of human intelligence, with its results published in a form that is useful and comprehensible to educators at all levels and does not lend itself to the traditional bureaucratic abuse to which the I.Q. concept has been subjected"[25].

Application of Delphi for a Firm

The advantages for a firm in the application of Delphi lies in the insight, which the firm will gain into the general environment, into an industrial sector and into its own structure.

Insight into the general environment of a firm and into an industrial sector

The insight into the general environment of the enterprise is obtained by an insight into the probable characteristics of the world of the future or into more specific problems of that future world, such as educational innovations, technological and scientific developments and their societal consequences, etc.

Insight into the firm itself

We have seen that the methods cited are of assistance in the determination of events (technological and social occurrences). Given the events, the enterprise has to state its goals in order to determine its policies. Insight for an enterprise either into its goal structure (hierarchy of goals and sub-goals, goals oriented planning) or into the policies to be followed (alternative applications of resources and actions) is not always clear.

In some enterprises in the United States and in Western Europe, Delphi has been used for this purpose[26].

It is possible to view the problem more narrowly, viz., for a firm producing only one product. Chambers, Mullick and Smith, through their experiences in

Corning Glass Works, have studied the application of Delphi, besides other forecasting techniques, for the forecasting of the life cycle of a product[27].

Generally speaking, a firm is a multi-product producer. Strategy then goes beyond the life cycle of products and is linked up with the vitality of the enterprise itself and not with that of certain well defined products.

This application inside the firm can be used to demonstrate the advantages of the Delphi technique with regard to the convergence of opinions. The following rules apply:

1) the steering group for the Delphi exercise in the enterprise consists of the systems analysis group or of a special group created in the enterprise especially for the occasion;

2) the persons contacted for the Delphi exercise are individuals with very good insight into the enterprise but who are not necessarily the most important people in the enterprise. The participants are chosen from all divisions of the enterprise (e. g., production, sales, personnel, finance). It is preferable for the participants to be chosen and the directives for Delphi to be given by outside advisers;

3) it is possible for the steering group first to take the initiative of supplying a *Strengths-Weaknesses-Analysis* of the enterprise and, if the participants have enough spare time, of holding a brain-storming exercise. At that moment, brain-storming will form an exercise in prospective thinking concerning alternative goals and alternative policies or strategies for the enterprise.

   Evaluation of the alternative goals and policies of the enterprise are, however, only possible using Delphi techniques;

4) in order not to waste the participants' time, the questionnaire approach of Delphi in several rounds with feedback of the answers concerning goals, goal structure and alternative policies is used ;

5) the Delphi exercise is repeated at each level of decentralization of the enterprise with regard to the goals of each enterprise level, the place of the decentralized goals in the hierarchical goal structure and the decentralized alternative policies.

Participants in such a Delphi exercise in an enterprise are always very enthusiastic as they participate in this way in the preparation of decisions, which will also determine their future (a form of *Participatory Planning*). Because of this involvement the tasks will be executed in a better way than before. Finally the decision makers of the enterprise have to be convinced about the outcome.

Applications of Delphi for the Public Sector

Of course, there are all the applications, which were already mentioned, in the interests of the welfare and well-being of the population. In the strict sense of the public sector however, studies in education[28] and defense can be mentioned as examples.

As far as defense is concerned, studies on the uses of the A-bomb by the Soviets, and the civil defense study were already mentioned. In the long-range forecasting work of 1963-64 by Gordon and Helmer, two of the six areas were concerned with the probability and prevention of war and with future weapon systems[29]. Delphi can be extremely useful for decision making in defense in connection with the determination of alternative futures and for the definition of alternative strategies (attacks or responses).

Some weak points of Delphi

1) The Delphi method may be a time, energy and money consuming business and for that reason *Nominal Group Technique* may be a correct first approximation, or even a substitute?

2) The fact that mutual connections between the events are not taken into consideration forms another weak point for Delphi. *Cross-Impact* using Delphi as a starting point take these interconnections of events into consideration.

1.4 NOMINAL GROUP TECHNIQUE

General Description

Similar to the Delphi method, the Nominal Group Technique is a method for obtaining and processing judgmental data. A group of especially knowledgeable individuals (experts), representing all stakeholders, is formed, which comes together in a closed meeting, at least during a half day. Out of experience the group is composed of at least fifteen individuals. A steering panel or a panel leader leads the group.

The nominal group technique consists of a sequence of steps, each of which has been designed to achieve a specific purpose[30].

1) The steering group or the panel leader carefully phrases as a question the problem to be researched. Much of the success of the technique hinges around a well-phrased question. Otherwise the exercise can easily yield a collection of truisms and obvious statements. A successful question is

quite specific and refers to real problems. The question has to have a singular meaning and a quantitative form as much as possible.

2) The steering group or the panel leader explains the technique to the audience. This group of participants is asked to generate and write down ideas about the problem under examination. These ideas too have to have a singular meaning and a quantitative form as much as possible. Participants do not discuss their ideas with each other at this stage. This stage lasts between five and twenty minutes.

3) Each person in round-robin fashion produces one idea from his own list and eventually gives further details. Other rounds are organized until all ideas are recorded.

4) The steering group or the panel leader will discuss with the participants the overlapping of the ideas and the final wording of the ideas.

5) The nominal voting consists of the selection of priorities, rating by each participant separately, while the outcome is the totality of the individual votes. A usual procedure consists of the choice by each participant of the *n* best ideas from his point of view, with the best idea receiving *n* points and the lowest one point. All the points of the group are added up. A ranking is the democratic result for the whole group.

6) Out of experience, one may say that there is still much dreaming, even between experts. Therefore the group was also questioned about the probability of occurrence of the event. In this way they became more critical even about their own ideas. The probability of the group is found as the median of the individual probabilities.

7) Finally, the group rating *(R)* is multiplied with the group probability *(P)* in order to obtain the effectiveness rate of the event *(E)*:

$$R \, x \, P = E$$

Once again, the effectiveness rates of the group are ordered by ranking.

Advantages and Disadvantages of the Nominal Group Technique

The Nominal Group Technique is faster and less costly than Delphi. Contrary to Delphi, convergence is not aimed at, but final voting is used. In this way, Nominal Group Technique is advisable for a quick result or as a preliminary version of Delphi.

Some Applications

We have conducted many experiments in nominal group technique over the years. The experiments cover the period from 1977 until now. The first experiment took place during a seminar for managers led by us in Johannesburg, South Africa on June 14, 1977, in collaboration with the School of Business Leadership of the University of South Africa. Twenty-seven participants listed some thirty-two events. These events would have to be included in a scenario on the future of South Africa in response to the following question: what are the most important factors to be included in a scenario on the future business outlook (5-10 years) in South Africa[31]?

It was good judgment that this panel of business managers ranked first two political and non-economic factors before the economic ones.

Over the years, many Nominal Group Technique exercises were conducted as seminars in several university institutions[32].

In 1986, a Nominal Group Technique was experimented by a group of 23 postgraduate students from French-speaking Africa in the Institute for Development Policy and Management of the University of Antwerp. The question was: which events will most influence the economic outlook of Developing Countries in the following ten years? (See Annex 1 of this chapter). The reader himself may judge about wisdom of the students' judgments. Anyway, a *Post Evaluation* of the economic outlook of Developing Countries over the period 1987-1996 would show an amazing approximation of the real outcome over that period.

Another experiment took place in Lithuania on November 9, 2001. The group was composed of 18 graduate students chosen at random from the faculties of business management and engineering of the Vilnius Technical University of Lithuania (See Annex 2 of this chapter).

Finally, the Nominal Group Technique was applied for the Facilities Management sector of Lithuania on October 15, 2002. Fifteen delegates from the facilities sector, from the ministerial departments and from the academic world composed the group of experts (See Annex 3 of this chapter).

In all the experiments, the criteria of Delphi were applied. In the South African and in the Facilities Sector exercises the participants were experts. The other experiments with students did not live up to the expert criterion. The students were chosen only because other candidates were not available.

The disadvantage of the nominal group technique is that no convergence is reached between all participants, but that only a majority voting is obtained, though corrected by a probability factor. The Delphi method on the contrary stresses an opinion convergence.

1.5 THE CROSS-IMPACT TECHNIQUE

Gordon and Helmer found cross-impact relations for Kaiser Aluminum and Chemical Corporation in 1966[33]. The real first publication on the new method dates from 1968[34]. Since then, considerable improvements have been made in the technique by the staff of the Institute for the Future, especially by S. Enzer[35]. He defines cross-impact as follows:

> "Cross-impact analysis is an experimental tool that attempts to evaluate average likelihoods of occurrence for each event in a set of interrelated events, considering possible sequences and occurrences or non-occurrences among the events in the set"[36].

The following steps are taken:

1) the events are found after several Delphi rounds;

2) the Delphi group is asked for the likelihood of occurrence of each event. The likelihood of occurrence of each event used finally is the median of a final round (*Initial Probabilities*);

3) each event is matched by the panel to each other event in the cross-impact matrix. Each entry in the matrix shows the new likelihood of occurrence of an event if another event, occurs or does not occur (*Conditional Probabilities*);
In its bilateral fixing of the probabilities the panel could be illogical when the matrix is viewed as a whole. A computer will indicate - given the initial probabilities - the logical limits between which the conditional probabilities can vary. The panel's median values must be situated within these limits;

4) until now only the influence of one single event on another single event was studied. Not only several events will influence the occurrence or non-occurrence of other events, but even the sequence of the several events is very important. This sequence of events is considered either voluntaristic or stochastic.
It is also possible to introduce the influence of policies or actions on the likelihood of occurrence of an event;

5) once the sequence of events is determined, the computer will perform an operation, which is comparable to casting dice, in order to define the *Final Probability* of occurrence of each event.

The computer, which was used at the "Institute for the Future" could run a cross-impact matrix up to a 20 x 20 array.

The main advantage of cross-impact above the ordinary Delphi technique is that Delphi only proceeds event by event, while cross-impact takes into consideration the interactions between the possible events.

Once again we go back to the first experiments of cross-impact. The cross-impact technique was used for exploration of the economic, social and political environment of Europe for the decade 1970-80[37]. Other applications have been: a study on alternate futures for education[38], the study of urbanization over the period 1970-85 at the Seminar on Public Administration at the College of Europe in Bruges (Belgium)[39], the study on the future of the Canadian economic, social and physical environment[40] etc. The test carried out by Gordon and Hayward in 1968 was based upon the individual recollections of 1950 of specialized officers and was concerned with a military matter, viz., the estimation in 1950 of the possibility of realizing the Minuteman and Polaris rockets [41].

In the next chapter we tackle all kinds of simulation methods, which are mostly treated nominally.

CHAPTER 1 - ANNEX 1

A NOMINAL GROUP TECHNIQUE ON THE ECONOMIC SITUATION OF THE DEVELOPING COUNTRIES OVER THE PERIOD 1987-1996

WHICH EVENTS WILL THE MOST INFLUENCE THE ECONOMIC OUTLOOK OF THE DEVELOPING COUNTRIES IN THE FOLLOWING TEN YEARS?

Institute for Development Policy and Management
of the University of Antwerp
March 1986

53 events were recorded of which 12 got no points

23 postgraduate students from French speaking Africa participated

The following ranking was obtained for the most important events (Table1A1.1)

Table 1A1.1 Most Important Events influencing the Developing Countries

Events	R	Rank	P	E = R x P	Final Rank
1. Complete refusal to pay their external debt by Developing Countries	25	3	0.6	15	1
2. External debt of African countries exceeds 30% of exports turnover	22	5	0.6	13.2	2
3. At least 30% of the machines in developed countries will be completely automated	25	3	0.5	12.5	3
4. At least 15% increase in regional warfare	18	6	0.5	9	4
5. At least 10% increase in external debt of developing countries	11	13	0.8	8.8	5
6. At least 70% increase in Arab religious fanaticism will increase political tensions in the world	17	8	0.5	8.5	6
7. At least 10% increase of famine in the poorest developing countries	14	9	0.6	8.4	7
7bis. External debt will reach 60% of the export turn-over in developing countries	14	9	0.6	8.4	7
9. At least 10% increase in unemployment in developing countries and at least 5% in developed countries	10	14	0.8	8	9
10. Decrease by 50% of crude oil prices	33	1	0.2	6.6	10
11.The Gross National Product per head in developed countries will be at least 130% more than in developing countries	13	12	0.5	6.5	11
12 Race in nuclear weapons possession of the large powers, but without an atomic war	28	2	0.2	5.6	12
13. Probability of a World War is only 40%	18	6	0.3	5.4	13
14. Increase by at least 50% of pollution in the world	9	15	0.5	4.5	14
15. Terrorist attacks increase by at least 10%	7	16	0.6	4.2	15
15bis. Probability of 0.7 that white domination will completely disappear in South Africa	7	16	0.6	4.2	15

The results show the importance of the probability of realization of the events. Events 1, 2, 4, 5, 6, 7bis, 9 etc. improve their position in a significant way. On the contrary, event 10 moves from the first rank to the tenth one and event 12 goes down from the second to the twelfth rank.

CHAPTER 1 - ANNEX 2

APPLICATION FOR THE ECONOMIC SITUATION OF LITHUANIA OVER THE PERIOD (2002–2011)[42]

Which events (maximum 5 per participant) will most influence the economic outlook of Lithuania in the following ten years (economic, technical, political, social, medical, and other events)?

November 9, 2001

23 events were proposed

2 events got no points by 18 graduate students chosen at random as experts

The steering group was composed of W.K. Brauers and

PhD student N. Lepkova from Vilnius Gediminas Technical University

The following final ranking was obtained (Table 1A2.1).

Table 1A2.1 Most Important Events influencing the economic outlook of Lithuania in 2002-2011

No	Events	Total	Rank
1.	Member of European Union	18	1
2.	Rise of Information Technology	15	2
3.	Privatization of Education and of other Public Services	12	3
4.	Member of NATO	11.25	4
5.	Introduction of Progressive Taxation	9.5	5

Due to lack of time, the other events were not taken into consideration for a final ranking. As they all got preliminary points, their influence has also to be considered (Table 1A2.2).

Table 1A2.2 Complete List of the Most Important Events influencing the Economic Outlook for Lithuania in 2002-2011

	Events	Total points **R**	Rank	Median of probabilities ***P***	***E*** = ***R***×***P***	Final Rank
1.	Internal Security	2	19			
2.	Member of NATO	15	5	0.75	11.25	4
3.	No changes in Economic Policy	0				
4.	Higher Quality of Education	11	7			
5.	Rise of Information Technology	20	1	0.75	15	2
6.	Important Foreign Investments	14	6			
7.	Introduction of Progressive Taxation	19	2	0.50	9.5	5
8.	Possibility to buy Private Property	7	12			
9.	Member of European Union	18	3	1	18	1
10.	General Use of E-business	8	10			
11.	Privatization of Education and of other Public Services	16	4	0.75	12	3
12.	Technological Progress	10	11			
13.	Professional Army instead of Draftees	8	10			
14.	Young People in Politics	2	19			
15.	Liberal Economy	0				
16.	More Advanced Technology in Medicine	8	10			
17.	Effective Social Policy in E.U.	7	13			
18.	Large Increase in Exportation	6	14			
19.	Important Political Influence in the Economy	5	14			
20.	Influence of Russian Economy	4	16			
21.	Closing down of Atomic Plants	3	18			
22.	Skillful and High Moral Politicians	11	7			
23.	Higher Birth Rate	4	16			

CHAPTER 1 - ANNEX 3

NOMINAL GROUP TECHNIQUE ON THE BUSINESS OUTLOOK OF THE FACILITIES SECTOR OF LITHUANIA OVER THE PERIOD (2003–2012)[43]

Which events will most influence the business outlook of the facilities sector of Lithuania in the following ten years (economic, technical, political, social, medical, and other events)?

October 15, 2002

Definition of Facilities Sector in Lithuania

The Facilities sector in Lithuania provides the following services:

- Acquisition, leasing and renting of existing buildings;
- Management of buildings, which is a multifunctional service. This means that all supervision, maintenance and repairing is included in the sector.

The Facilities Sector is only a very small sector in Lithuania, composed of a small number of small firms, which even perform other tasks outside facilities management, such as waste management. The largest firm in the sector counts only 179 employees. Official statistics are not separately available for the Facilities Sector.

In theory the facilities sector could include the entire management of Corporate Real Estate. This means the effective management, which is called the *Fifth Resource*. Indeed, in the report of "The Industrial Development Research Foundation of the United States" the corporate real estate assets are indicated as a *fifth resource*, after the resources of people, technology, information and capital[44].

Preliminary Information by the Panel Leader

The panel leader Prof. Dr. W.K. Brauers gave some preliminary information in order to assist the participants in their judgments:

1) The events that were proposed on the general outlook of Lithuania over the period 2002-2011 (see this Chapter Annex 2)

2) Indicators of Lithuania towards membership of European Union (Table 1A3.1)

Table 1A3.1. Main indicators concerning Lithuania for the year 2001[a)]

Attributes	Eur. Union	Lithuania	Last Country	
GDP/Capita (in Euro)	23,200	8,700	Latvia	7,700
Δ % GDP	1.7	5.9	Malta	–0.8
Unemployment %	7.6	16.5	Slovakia	19.4
Inflation %	2 (2.3)	3.3	Hungary	12.4
Exports to E.U. (%)	–	47.8	Malta	41.3
Imports from E.U. (%)	–	44.0	Lithuania	
Agriculture (% of GDP)[b)]		7.0	Lithuania	
Agriculture (% of Employment)[b)]		16.5	Poland	19.2

[a)] Source: Commission of the European Union. Report about the ten countries, candidate members of the European Union. The figures are standardised after the European Union norms.
[b)] High figures in agriculture mark a country as being less developed, certainly when the % of employment in agriculture is high.
A low productivity in agriculture occurs if the percentage of agriculture in GDP is smaller than the percentage of employment.
In Poland productivity in agriculture is extremely low. With 19,2% of the population only 3,4% of GDP is realized in agriculture.

3) Information about Technological Forecasting

Technological Forecasting stands for the introduction or generalization of new products or materials in the future. Examples could be the general use of solar energy and of combustion cells in households and office buildings.

4) Information about SWOT analysis

SWOT analysis for the sector under focus consists in identifying possible Strengths, Weaknesses, Opportunities and Threats.

The low salaries in Lithuania could be a source of Strength. On the contrary, low productivity could be a Weakness.

Opportunities could come from expansion in the European Union. Indeed, in some West-European countries the facilities sector is still heavily decentralized and in addition perhaps underestimated. The experience of Lithuania in this field could be useful. Threats may come from the danger of war.

Nominal Group Technique for the Facilities Sector in Lithuania

1) Composition of the Group

The group of especially knowledgeable people was composed of delegates from the facilities sector, from the ministerial departments and from the academic world (15 participants).

The steering group was composed of Prof. Dr. W.K. Brauers and PhD student N.Lepkova. The panel leader directing the exercise was Prof.Dr. W.K.Brauers.

Each participant has chosen the most important five events from his point of view, with the most important event receiving five points and the less important event one point. Of course, the members of the steering group did not participate in the voting.

2) The Outcome (see table 1A3.2)

Table 1A3.2. Important Events influencing the Business Outlook of the Facilities Sector of Lithuania over the period 2003-2012 (ranked by importance)

	Events 2003-2012	Given Points **R**	Rank	Median Probabilities **P**	**E=RxP**	Final rank
1	Member of European Union	37	1	0.75	27.75	1
2	Large increase in foreign capital	20	2	0.75	15	2
3	More competition between facilities management companies	16	3	0.88	14.08	3
4	Large increase in GDP	16	3	0.75	12	4
5	New materials and technologies	12	6	0.75	9	5
6	Stability in international security	14	5	0.50	7	6
7	Higher quality in building construction	8	11	0.75	6	7
8	Application of new information technologies to facilities management	9	9	0.63	5.67	8
8	More relations with foreign companies having more experience in facilities management	9	9	0.63	5.67	8
10	Better legislation in supervision sector	11	7	0.5	5.5	10
11	Optimal quality-price relation for services	7	13	0.75	5.25	11
12	Better public estimation for facilities management	8	11	0.63	5.04	12
13	Increase of the cost of living	10	8	0.50	5	13
14	Positive influence of the laws in favor of facilities management	6	16	0.75	4.5	14
15	Higher qualification of the staff in facilities management	5	20	0.75	3.75	15
16	Change taxation for buying and selling property	7	13	0.50	3.5	16
17	Higher demand for construction	6	16	0.50	3	17
17	Higher level of education	6	16	0.50	3	17
19	International exchanges between students in facilities management	7	13	0.25	1.75	19
20	Increase of industrial sector and decrease of agricultural sector	6	16	0.25	1.5	20
21	Improvement of sanitary services	4	21	0.25	1	21
22	Increase of individual property of housing	1	22	0.25	0.25	22
	Total Points	**225**			**145.21**	

The introduction of probabilities of realization, introducing a sense of reality and presenting a guaranty against wishful thinking, produces quite some changes in the ranking.

The total 225 is a control figure for the group result. Indeed, each participant could distribute maximum: 5+4+3+2+1 = 15 points. With 15 participants, the total has to be not more than 225. It could be less, as each participant is not obliged to allot 15 points. The total of the given points, here namely 225, means that each participant used his rights completely. The reality check, however, diminishes the figure to 145.21.

The results of the Nominal Group Technique by a group of especially knowledgeable persons around the Facilities Sector are extremely interesting.

Nineteen % of all the voting points go to the Membership of the European Union, which indeed will not only influence the Business Outlook of the Facilities Sector, but the Economic Outlook of Lithuania itself.

Concerning international trade, maybe not so much will change for the ten new members of the European Union. Indeed, they were already part of a European Free Trade Zone with a large possibility for European international trade. Lithuania may form an exception, as being until now from the ten new members the less dependent for trade with the European Union (see table 1A3.1).

More important for the ten new member countries will be that in the long run more and more interest will be shown for investment in these countries. Therefore, the experts of the Nominal Group Technique could be right to rank second a large increase in foreign capital. A positive consequence could be a large increase in Gross Domestic Product (rank 4), together with a negative one: an increase in the cost of living (rank 13). This last consequence could look strange. It is interesting to develop this point a bit further.

When countries with a lower productivity join an economic union, the productivity of the international tradable goods sectors, mostly the industrial sectors, will rise over time to approach as much as possible the higher productivity of the other countries. However, the "Balassa–Samuelson Effect" maintains that this increase is less the case for the nontradables[45]. The international tradable goods sectors will have the opportunity to increase their wages. By osmosis of the labor market, the nontradable sectors have to raise their wages too, although their productivity does not increase in equal importance. Inflation is the result with an increase in the cost of living.

In particular, the facilities sector, belonging to the nontradables, will have to pay higher wages without a proportional increase in production. Will this explain more competition between facilities management companies (rank 3)?

Furthermore, the facilities sector is aware of its other problems in the future:

- new materials and technologies
- higher quality in building construction

- application of new information technologies
- more relations with foreign companies having more experience in facilities management (does this mean possible mergers?)
- better legislation in supervision sector
- optimal quality–price relation for services
- better public estimation for facilities management
- positive influence of the laws in favor of facilities management
- higher qualification of the staff
- a change in taxation for buying and selling property
- higher demand for new construction
- international exchange of students in facilities management, etc.

NOTES PART 2 CHAPTER 1

[1] *Images du monde en l'an 2000*, Centre européen de coordination de recherche et de documentation en sciences sociales, Vienna, 1976.

[2] E. S. Quade, *Cost-Effectiveness: Some Trends in Analysis*, Rand Corporation, P-3529-1, Santa Monica (CAL), March 1970, 9-10.

[3] E. Jantsch, *Technological Forecasting in Perspective*, OECD, Paris, 1967, 136.

[4] E. Jantsch, op.cit.136.

[5] E. S. Quade; W. I. Boucher, *Systems Analysis and Policy Planning: Applications in Defense*, Elsevier, New York, 1968, 334.

[6] N. Dalkey; O. Helmer, *An Experimental Application of the Delphi Method to the use of Experts*, Management Science, April 1963, 458-487.

[7] E. S.Quade, *op. cit.*, 10.

[8] N. Dalkey; O. Helmer, *op. cit.* 458.

[9] T. J. Gordon and 0.Helmer, *Report on a Long-Range Forecasting Study*, P-2982, Rand Corporation, Santa Monica, Cal. September 1964. O. Helmer, *Social Technology*, Basic Books, Inc. New York and London, 1966, 44-96. D. L. Schmidt, *Creativity in Industrial Engineering*, P-4601, Rand Corporation, Santa Monica Cal. March 1971.

[10] 0. Helmer, *The Use of the Delphi Technique in Problems of Educational Innovations*, P-3499, Rand Corporation, Santa Monica, Cal. December 1966.

[11] T. J. Gordon, D. L. Little, H. L. Strudler, D. D. Lustgarten, *A Forecast of the Interaction between Business and Society in the Next Five Years*, R-21, Institute for the Future (I. F. F. Middletown, Conn. April 1971.

[12] T. J. Gordon and R. H. Ament, *Forecasts of Some Technological and Scientific Developments and Their Societal Consequences,* R-6, I.F.F. Middletown, Conn. September 1969. R. H. Ament, *Comparison of Delphi Forecasting Studies in 1964 and 1969*, Futures, Vol. 2, N' 1, 1970, 35-44. S.Enzer, *Some Developments in Plastics and Competing Materials by 1985*, R-17, Institute for the Future, Middletown, Conn. January 1971.

[13] R. De Brigard and 0.Helmer, *Some Potential Societal Developments 1970-2000*, R-7, I. F. F., Middletown, Conn. April 1970. 29.

[14] T. J. Gordon, *A Study of Potential Changes in Employee Benefits*, 3 Vol., I.F.F. Middletown Conn. April 1969.

[15] 0.Helmer, T. J. Gordon, S. Enzer, R. De Brigard, R. Rochberg, *Development of Long-Range Forecasting Methods for Connecticut.* A Summary, R-5, I.F.F. Middletown, Conn. September 1969. S.Enzer and R.De Brigard, *Issues and Opportunities in the State of Connecticut: 1970-2000*, R-8, I. F. F. Middletown, Conn. March 1970.

[16] H.S. Becker, *A Method of Obtaining Forecasts for Long-Range Aerospace Program Planning*, WP-7, I.F.F. Middletown, Conn. April 1970.

[17] 0. and H. Helmer, *Future Opportunities for Foundation Support*, R-11, I.F.F. Middletown, Conn. June 1970.

[18] See for instance N.C. Dalkey, *Quality of Life*, p-3805, Rand Corporation, Santa Monica, CAL. March 1968. N.Rescher, *Delphi and Values*, p-4182, Rand Corporation, Santa Monica, CAL. September 1969, N.C. Dalkey and D.L. Rourke, *Experimental Assessment of Delphi Procedures with Group Value Judgments*, ARPA order N'189-1, Rand Corporation, Santa Monica, CAL. February 1971.

[19] E.W. Paxson, *A Delphi Examination of Civil Defense*, 2 vol. prepared for ARPA, Order N°189.1, Rand Corporation, Santa Monica, CAL. May 1970.

[20] N.C.Dalkey, R.Lewis and D.Snyder, *Measurement and Analysis of the Quality of Life; with Exploratory Illustrations of Applications to Career and Transportation Choices*, prepared for U.S. Department of Transportation, R.M. 6228-DOT, Rand Corporation, Santa Monica, CAL. August 1970.

[21] 0. and H.Helmer, *Future Opportunities for Foundation Support*, op. cit.

[22] At this stage, brainstorming is advisable.

[23] 0. and H.Helmer, *op. cit.*, 10.

[24] *Op. cit.*, 13.

[25] *Op. cit.*, 21.

[26] Many of these applications are not communicated to the outside world by the enterprises. The first application known is that of T.R.W. in the U. S. See: H. Q. North and D.L. Pyke, *Probes of the technological future. The Corporate planner should pursue new means of technological forecasting to ensure his company's share in tomorrow's markets*" Harvard Business Review, May-June 1969, 68-82.
For Western Europe, we cite the example of the multinational photographic group Agfa-Gevaert. See: D. Oates, *Focus on the Future at Agfa-Gevaert, European photographic group aims to clarify its objectives*, International Management, August 1972, 23-26.

[27] J.C. Chambers, S. K. Mullick and D. D. Smith, *How to choose the right forecasting technique*, Harvard Business Review, July-August 1971, 45-74.

[28] 0.Helmer, *The Use of the Delphi Technique in Problems of Educational Innovations*, op. cit.

[29] T.J. Gordon and 0.Helmer, *Report on a Long-Range Forecasting Study*, op. cit.

[30] Brauers developed the approach of the Nominal Group Technique, which is explained here (W.K.Brauers, *Nominal Methods in Group Multiple Decision Making*, Research Paper N°3, Institute for Developing Countries, University of Antwerp, RUCA, Antwerpen, 1987).
The Nominal Group Technique was first elaborated at the University of Wisconsin by Delbecq and Van De Ven: A.H.Van De Ven, A.L.Delbecq, *Nominal versus Interacting Group Processes for Committee Decision Making Effectiveness*, Academy of Management Journal, vol. 14, N°2, June 1971, 203 and fol.
A.L.Delbecq , A.H.Van De Ven, D.H.Gustafson. *Group Techniques for Program Planning: A Guide to Nominal Group and Delphi Processes*, Scott, Foresman and Co, Glenview (Ill.) 1975.

S.L.Gill, A L. Delbecq and T. J. Snodgrass, three chapters on Program Planning, Nominal Group Technique and Value Engineering, in Shirley Jones (Ed.), *Group Process Planning and Problem Solving Methods*, N.Y., John Wiley and Sons, 1981.
A L. Delbecq, "The Nominal Group as a Technique for Understanding the Qualitative Dimension of Client Needs," in *Assessing Health and Human Service Needs*, R.A.Bell, editor, N. Y., Human Services Press, 1983, 210-218.
A.L.Delbecq, A.Van de Ven and D.Gustafson, *Group Techniques for Program Planning*, 2nd edition, Greenbriar Press, Madison, Wisconsin, 1986.

[31] W.K.Brauers, *Nominal Group Technique and Cross-Impact Analysis*, University of Antwerp (RUCA), Faculty of Applied Economics, September, 1978.

[32] W.K.Brauers, *Nominal Group Technique and Cross-Impact Analysis*, op. cit.
W.K.Brauers, *Het bedrijfsleven in de jaren 80*, HIBO, Gent, March 1980.

[33] 0.Helmer, *Report on the Future of the Future-State-of-the-Union Reports*, Institute for the Future, Middletown, Conn. R-14. October 1970, 30.

[34] T.J. Gordon and H. Hayward, *Initial Experiments with the Gross Impact Matrix Method of Forecasting*, Futures, December 1968, 100 - 116.

[35] S. Enzer, W.I. Boucher and F.D. Lazar, *Futures Research as an Aid to Government Planning in Canada*: Four Workshop Demonstrations R-22, 1. F. F. Middletown, Conn. August 1971.

[36] S. Enzer, *Delphi and Cross-Impact Techniques, An Effective Combination for Systematic Future Analysis*, Institute for the Future, WP-8, Middle- town, Conn. June 1970, 6.

[37] S. Enzer, *A Case Study using Forecasting as a Decisionmaking Aid*, WP-2, I.F.F. Middletown Conn. December 1969. S. Enzer, Delphi and Gross-Impact Techniques an Effective Combination for Systematic Futures Analysis, WP-8, I.F.F. Middletown Conn. June 1970.

[38] T.J. Gordon, *Cross-Impact Matrices. An Illustration of Their Use for Policy Analysis*, Futures, December 1969, 527-531. R. Rochberg, T.J. Gordon, 0.Helmer, *The Use of Gross-Impact Matrices for Forecasting and Planning*, R-10, 1. F. F. Conn. April 1970. This study is also an illustration of how sensitivity analysis is afterwards used when a model has been set up, namely, the cross-impact model.

[39] R.M. Lees and others, *New Techniques in Public Administration*, Vol. 1, College of Europe, Bruges 1971, 15.

[40] S. Enzer, W.L. Boucher and F.D. Lazar, *op. cit.*

[41] T.J. Gordon and H. Hayward, *op. cit.*

[42] W.K.Brauers, N.Lepkova, *The application of the Nominal Group Technique to the Economic Outlook of Lithuania over the period 2002-2011*, UKIO technologinis ir ekonominis vystymas (technological and economic development of the economy) Vilnius Gediminas Technical University, Lithuania, VIII, N°1, 2002,19-24.

[43] W.K.Brauers, N.Lepkova, *The application of the Nominal Group Technique to the Business Outlook of the Facilities Sector of Lithuania over the period 2003-2012*, International Journal of Strategic Property Management, Vilnius Gediminas Technical University, Lithuania and Napier University, Scotland, vol.7, N°1, 2003, 1-9.

[44] McGregor,W.; Shiem-Shim Then,D. *Facilities management and the business of space*. Butterworth-Heinemann, London, 1999.

[45] P.A.Samuelson *Theoretical Notes on Trade Problems*, Review of Economics and Statistics, 1964, 145-154.

B.Balassa *The Purchasing Power Parity Doctrine: A Reappraisal*, Journal of Political Economy, 1964, 584-596.

P.A.Samuelson. *Facets of Balassa-Samuelson Thirty Years Later*, Review of International Economics, 1994, 201-226.

E.Balazs, I.Drine, K.Lommatzsch, C.Rault. *The Balassa-Samuelson Effect in Central and Eastern Europe: Myth or Reality?* Working Paper 483, The William Davidson Institute, University of Michigan Business School, July 2002.

CHAPTER 2

SIMULATION AND SENSITIVITY ANALYSIS

The search for analogy in a micro-world, in order to have a better insight in the real world, is essential to simulation. Gordon, Enzer and Rochberg gave in our view the most complete description of simulation[1].

2.1 CATEGORIES OF SIMULATION

1. Mechanical Analogs

Nowadays one has rather to speak of "electronic devices". A good example is a wind tunnel model of a plane.

The device used in the *Electronic-Mechanical Analog* is sometimes called a simulator, such as a flight-simulator for the trainee pilot. Consequently, the electronic-mechanical analog is much used by the airlines and the air force.

Another example comes from the textile industry. Fibers are tested on their strength and their sensitivity to light. On the one hand, the fibers are tested on their sensitivity by a computer controlled device. On the other hand, exposure to a strong artificial light, more intensive than extended exposure to daylight, controls the characteristic of sensitivity to light.

2. Metaphorical Analogs

The growth of a bacterial colony depicting human population growth is an example of a metaphorical analog. In the past, the behavior of recruits in the Belgian Army, who formed a closed group, was studied as a simulation for the attitudes of Belgian youth as a whole, for instance for the habit of smoking cigarettes.

This simulation differs from the method of *Polls*, as in the latter method the persons polled do not form a group but are rather dispersed, while in addition statistical laws are involved concerning the significance link between small and large numbers.

3. Game Analogs

In Game Analogs, interactions between players are taken to represent social interactions. In this sense, i.e., within the framework of a very broad definition of gaming, some consider theater acting, training of individuals, football and gambling as forms of gaming[2].

We should like to define gaming more narrowly, viz., as simulation with a verbal, qualitative, non-mathematical imitation of reality by people taking themselves seriously in an exercise especially drawn up for the occasion in which all parties take their own decisions (*homo ludens*).

As examples, we cite war games, diplomatic games, games concerning society and business games. In addition, the playing of children in their imitation of adults means a little world simulation of reality[3].

National defense has a long tradition of gaming in the form of war gaming or military gaming. In fact gaming originated in national defense.
Shubik pointed it out in the following way:

> "the overall trend in voluntary gaming in the last thirty years has been from an almost exclusive emphasis on military games to military-diplomatic games and to business games, and now more recently to games concerning society"[4].

What are the characteristics of gaming? Gaming could in fact learn very much from game theory if this theory could be developed in a more extensive way[5]. At present gaming is characterized by the following steps.

Before starting, very careful preparation of the game is necessary. The playing time is of course limited and after the play, the analysis must try to formulate the conclusions of the game.

Military games, being at the origin of gaming, produce a better insight for gaming in general. The following classes of military games have to be distinguished.

Closed or Open Games

In the closed game, the two parties, separated in different rooms, only communicate with each other through the members of a control group. The members of the control group are contacted by written reports, by telephone or by television screen. In the open game, the three groups are in one room and the papers and the maps of the adversary are directly seen. Open games are more useful for training and education purposes, the closed ones for tactical operations, reconnaissance, etc.

Closed games even more than open games belong to the group of Nominal Methods.

Manual or Computer Games

The games are either manual or computer assisted. This is in fact a characteristic of all kinds of simulation[6]. The military game is called fully automated if a robot-commander estimates in a fully automatic way the brigade's situation. Information of varying accuracy is given to him about the enemy' s battalions. The best intelligence capacity that he can possibly be given is the complete information about the enemy's battalions over, for example, the last twelve hours. Given this data, and without any cost considerations, he will act then in an optimal way.

The Level of Application

Military games are played at platoon, company, battalion, brigade and division level. If the game is played at force or service level or for defense as a whole, it becomes more a military-diplomatic game and will be based on a set of international milieu-studies or international diplomatic - political games[7].

Advantages and Disadvantages of Military Games for Decision Making

For other purposes, military games are educational:

1) they may help to discover alternative strategies to meet a certain future
2) they may even teach us about the desirability of a certain strategy.

According to Fisher, traditional military games are subject to severe constraints, viz.,

> "1) Unless the game is overly simplified (and hence perhaps essentially useless or misleading), one run-through takes an inordinately long time, and therefore is very expensive in terms of man-days of effort.
> 2) In a given "play" of the game, so many key factors vary that interpreting the significance of a single play has been most difficult. Repetitions, under controlled conditions, could help solve this problem; but in the past, the length of time required for one play made a large number of run-throughs infeasible.
> 3) It is difficult to make readily available to the players the large body of information (including a kit of analytical tools to utilize this information) required to permit them to assess adequately the relevant range of alternatives before deciding upon a move"[8].

To assist the information of the players of military games, and of computer assisted and fully automated ones in particular, the players are aided by all kind of models such as tactical models, support models and costs models. A good example for such an application is Project XRAY of RAND Corporation[9]. It concerns a game in which several game plays are possible in a relatively short period owing to:

1) open gaming under the form of an extension of Delphi;
2) assistance through an online time sharing computer system;

3) a memory, which is stored with large data banks of information and numerous submodel routines (modules);
4) one of these modules is a force - mix cost model with projections possible ten or more years into the future[10].

4. Mathematical Analogs

An example could be: a set of equations depicting the economic situation of a country.

We accept the various forms of simulation as given by Gordon, Enzer and Rochberg, with one exception, however: the mathematical analog, which should not be considered as a form of simulation. Mathematics as a deductive science is an instrument used by all other sciences. Every application of mathematics in other sciences would, according to the definition of the authors, represent a case of simulation. Such a definition of simulation would be too broad. In this context, we prefer to use the term "model" as defined in Part I, Chapter 2 of this book.

We shall narrow down the simulation of the fourth group to a much smaller class of phenomena. To the electronic-mechanical, the metaphorical and the game analogs we should like to add a fourth group of Analogs: *Parametric Analogs*.

2.2 PARAMETRIC ANALOGS

In parametric analogs, the starting point is formed in the first place, by fictitious parameters in a mathematical model, as statistical estimation may not be available while the parameters may be changed to find the outcomes of several situations[11].

Second, the starting point could be statistically estimated parameters anyway, but which are arbitrarily changed to find the outcome of several situations. If we spoke in Part I of this book about the shortcomings of econometric models, the use of parametric analogs may be helpful to soften this critique, for instance to cope with uncertainty.

All analogs, which were distinguished, fulfill the requirements described by Gordon, Enzer and Rochberg:

> "all have the ability to change with time and imposed conditions and are used where experimentation with an actual system is too costly, is morally impossible, or involves the study of problems which are so complex that analytic solution appears impractical"[12].

In fact, the parametric analog with changing parameters belongs to the field of sensitivity analysis, which we describe now.

2.3 SENSITIVITY ANALYSIS

Sensitivity analysis studies the very sensitive part of the variation pattern between an outcome and one of the key parameters determining this outcome.

Let us start with an example in order to illustrate the meaning of sensitivity analysis. Dropping a bomb from an assault plane on a certain target involves a cost. However, target location plays an important part. To locate the target within 400 - 200 feet may cost the same but location within 20 - 12 feet will be much more expensive. A graph with the cost of realization and the location as axes will show the cost curve as a sharply rising curve.

Another example is the decline in the demand for gasoline if the tax on gasoline is increased. In this way, it is possible to determine a sensitivity factor. The demand for gasoline changes in the opposite direction to the tax on gasoline. Moreover the sensitivity factor is much more negative, if the tax on gasoline rises from 300 to 400% of the factor cost, than a rise from 200 to 300%.

In Part I, 2.7 of this book we remarked that complete pollution abatement would be very difficult, unless one is not afraid of a prohibitive cost. Sensitivity analysis could estimate the sensitivity of pollution abatement to costs.

Concerning safety on the roads, prevention costs will rise exponentially in order to minimize the number of people killed on the roads. Costs could become prohibitive, whereas a portion of uncertainty remains. Complete safety is then an illusion.

Fisher puts uncertainty in the middle of the definition of sensitivity analysis:

> "suppose in a given analysis there are a few key parameters about which the analyst is very uncertain. Instead of using mean values for these parameters, the analyst may successively use several values (say, high, medium, and low) in an attempt to see how sensitive the results (the ranking of the alternatives being considered) are to variations in the uncertain parameters"[13].

Several values such as "high, medium, low" are also found in Scenario Writing. Scenario writing is, however, not so mathematically oriented as sensitivity analysis, although it may be based on mathematically inspired techniques.

NOTES PART 2 CHAPTER 2

[1] T.J.Gordon, S.Enzer and R.Rochberg, *An Experiment in Simulation Gaming for Social Policy Studies*, Technological Forecasting 1, 1970, 241 -261.

[2] M.Shubik, *On the Scope of Gaming*, The Rand Corporation p-4608, Santa Monica 1971, 6-8.

[3] Game theory is distinct from gaming. "It is a branch of mathematics which can be studied as such with no need to relate it to behavioral problems, to applications, or to games. In general game theoretic reasoning and analysis are of considerable use in constructing and analyzing games and gaming exercises constructed for any purpose. The solution concepts or theories offered in the theory of games may be regarded as normative or descriptive views of multiperson decision making ". M.Shubik, *On Gaming and Game Theory*, The Rand Corporation, P-4609, Santa Monica 1971, 3.

[4] M.Shubik, *On the Scope of Gaming*, op. cit. 3.

[5] Quade is very optimistic about the use of game theory for defense:

"game theory is now not only being successfully applied to tactical military problems - such as radar search and prediction, and the scheduling of missile launching under enemy pindown- but also to problems of budget allocation and force structure", E.S.Quade, *Cost- Effectiveness: Some Trends in Analysis*, The Rand Corporation, P-3529-1, March 1970, 5.

[6] The question is even posed whether on line or off line. See H.Sackman, *Advanced Research in Online Planning: Critique and Recommendations*, System Development Corporation, Santa Monica, 1970.

[7] During the cold war period, Swedish National Security was mainly based on a set of written international diplomatic-political games. In the Swedish Center for Research in Defense (FOA) military-diplomatic games were played. Swedish defense thinking did not favor Delphi techniques and cross-impact analysis, as it is feared that such applications would boil down to a unique solution. From our point of view, Delphi enlarges the horizon and cross-impact brings an insight into the most probable outcome. This does not mean that the most probable outcome has to be seen as the only possible outcome.

[8] G.H.Fisher, *Cost Considerations in Systems Analysis*, Elsevier, New York, 1971, 286-287.

[9] The Project XRAY is described by G. H. Fisher, *op. cit.* 286-301.

[10] RAND worked mainly for the American Air Force. Similar gaming exercises were developed for the Army by the American Army Staff and the Research Analysis Corporation.

[11] See as an example the economic integration of several national economies: W.K.Brauers, *Input-Output analyse en Internationale economische integratie*, Standaard Antwerpen - Utrecht 1968, 201-229.

[12] T.J.Gordon, S.Enzer and R.Rochberg, *op. cit.* p. 241.

[13] G.H.Fischer, *Cost Considerations in Systems Analysis*, 12.

CHAPTER 3

SCENARIO WRITING

What is the importance of scenario writing? Scenario writing provides the opportunity to compare several alternative actions, events or economic policies. Moreover, scenario writing tries to stress uncertainty.

3.1 ORIGIN AND CLASSIFICATION

In 1956, the oil companies could use extensive econometric models, but the closing down of the Suez Canal by President Nasser in July was not predictable in the models. The companies had to spend billions of dollars for a new transportation system. In order to pass the Cape of Good Hope, giant super tankers were created. Consequently, the oil companies looked for a device to replace the econometric models. As a reaction, this device had, on the one hand to be simple and on the other hand to present some guaranty for uncertainty. They found it in scenario writing as developed during the war by the RAND Corporation, the think tank in Santa Monica. This experience of the oil companies also proved to be very valuable during the later oil crises.

Nowadays scenario writing is in general use by many huge transnational companies. Beside these *Industrial Scenarios*, *Global Scenarios* rather deal with global economic issues[1].

Individual and collective scenario writing is still another classification. Individual scenario writing is a form of *Genius Forecasting*. Genius Forecasting is much criticized, as a single individual does not possess sufficient expert knowledge to cover every aspect of broad issues. Collective scenario writing on the contrary is a collective work, for instance of economists, sociologists, psychologists, engineers and researchers of the exact sciences.

3.2 HOW TO WRITE SCENARIOS?

The Preparation

As for every systems approach the following phases form the preparation to Scenario Writing.

An Inventory of the Existing Conditions

It is advisable that a *Swot-Analysis* accompanies this inventory i.e., an analysis of the Strengths, Weaknesses, Opportunities and Threats of the organization for which the scenario writing is made.

A Statement of the Aims of an Organization and Creative Thinking about Solutions towards these Aims

At this stage, Brainstorming, Nominal Group Techniques and Delphi are helpful techniques.

Some Precepts for Scenario Writing

1) The set of scenarios has to cover the whole *Statistical Universe* or a partition of that universe. We recall that by Statistical Universe we mean the set of all events.

> Let us consider the price setting of an oil company being a subset of the statistical universe. Two scenarios are foreseen. The first one supposes that prices of gasoline follow the usual seasonal trends and the other fluctuations of supply and demand. The second scenario assumes that all of a sudden a huge price increase of crude oil is the result of outside events, such as a unilateral decision of OPEC. From the past, we also recall the consequences of the closing down of the Suez Canal and of the Gulf War on oil prices.
>
> A railway company may consider two scenarios for the supply of passenger trains. The first one considers the normal traffic with its peak hours and seasonal fluctuations. The second one supposes that all of a sudden more trains are needed. For instance, a sudden increase in demand due to good weather conditions at the coast could be the reason. A restaurant at the coast side would face two comparable scenarios. This characteristic of scenario writing can also be called: completeness or the exhaustion of all possibilities.
>
> Beside partition, some premises may limit this completeness, such as the necessary absence of a total war and the preservation of democracy and human rights.

2) Scenarios have to be: plausible, internally consistent, relevant and original[2].

> Internally consistent means that each scenario excludes the other ones. Ipso facto, a combination of different scenarios in a new scenario has to be impossible.

3) Scenarios are strictly advisory and not deterministic.

For Ringland, scenarios are a form of modeling of the future, as is the case with a wind tunnel to test airframes and cars"[3].

4) Only two to four scenarios are mostly considered.

Why this limitation in the number of scenarios? The human brain cannot grasp combinations of many items at the same time. The psychologist Miller maintains that the human brain can consider only seven items at the same time[4]. At this point computers are superior, as they are better in checking inconsistency.

High, low and medium; optimistic, pessimistic and in between, are very often used to characterize scenarios. High and low are familiar since the *Growth-Share Matrix* in its measurement of growth and relative market share. One could say that the matrix suggests four scenarios: the well-known Stars (high growth and market share), Cash Cows (low growth and high market share), Problem Children (high growth and low market share) and Dogs (low growth and market share).

In 1993, three alternative scenarios predicted the macro-economic outlook of the Hokkaido region in Japan for the year 2000: an optimistic, a pessimistic and a medium one. It is interesting to note that the forecasts were made on the basis of input-output models[5]. This shows the possible symbiosis between several techniques, as was suggested earlier.

3.3 SOME WEAK POINTS OF SCENARIO WRITING

A weak point could be that a lack of interaction between the events characterizes the scenarios and therefore *Cross Impact* is more suitable.

It is also not advisable to look for probabilities of realization of each of the scenarios. Sometimes a panel of Delphi could be helpful for this point.

With Scenario Writing, and mostly with all techniques of *Creative and Prospective Thinking*, the problem of choice remains. They concern Value Judgments, whereby the ideas of an optimal solution and a second and a third best, are absent. In the next parts of the book, these shortcomings are discussed.

CHAPTER 3 – ANNEX 1

ECONOMIC POLICY IN TANZANIA WITH DIFFERENT INPUT-OUTPUT SCENARIOS AS A BASIS

The Input-Output method opens the way for setting up alternative structures, which could come into competition with one another. Each input-output structure would form the core of a "scenario"[6]. Economic policy has then to make a choice between the different scenarios.

The basis input-output table consists of an existing table, eventually updated, or is constructed for instance by benchmarking (similar countries) or by engineering data. The general well-being forms the final aim for this quantification.

Suppose a breakdown in following targets:

- maximization of a certain production X
- maximization of a certain production Y
- maximization of GNP
- minimization of the balance of payments deficit
- maximization of employment in man-years
- due to shortages:
  - minimization of a technical coefficient a_1
  - minimization of a technical coefficient a_2

The following scenarios for Tanzania were proposed[7] (World or African wars excluded).

SCENARIO I

Installation of a hydro-electrical power station and a steel factory

SCENARIO II

A new policy for agriculture and fisheries:

- the end of incentives to reduce agricultural production through price controls, export taxes, quotas, import subsidies;
- removing protection for non-agricultural products which increases the balance of payments deficit;
- liberalization of products such as coffee, cacao, cotton, sugar etc.

SCENARIO III

Installation of a new sugar factory and a cotton mill

SCENARIO IV

Intensified promotion of tourism together with a program for new roads

These scenarios have to correspond with the 3-Year Plan.

Too many input-output applications are only restricted to statistical work. More attention has to be given to the application for economic policy. Scenario writing, which forms a part of this model, is not new in combination with input-output.

Special care has to be taken for developing countries where more than elsewhere attention has to be given to benchmarking, engineering data and structural changes. The outcome will result in a choice of several alternative input-output structures under the form of several alternative scenarios.

CHAPTER 3 - ANNEX 2

SCENARIOS FOR LITHUANIA 2002 - 2011

Based on some interviews and the Nominal Group Technique, described in Annex 2 of Chapter 1 of this Part of the book, the following scenarios were deduced for the geopolitical and economic position of Lithuania in the period 2002-2011.

Vilnius, November 2001

Scenario I

Lithuania in the Period 2002 – 2011 becomes a Full-Member of the European Union and of NATO (World or European wars excluded)

The percentage of international trade with European Union countries, which was rather low in 2001, will go up (see Chapter 1, Table 1A3.1 of this Part II of the book).

Foreign investments will go up with an increasing rate.

Economic growth follows the world wide cyclical movement.

More privatization for public goods, public transport and education will occur.

In general, wages will go up. In industry, wages will go relatively more up than in the other sectors. Wages in government, relatively already high in the higher brackets, will relatively go down. Wages in primary, secondary and university education will relatively go up, with also better wages for researchers. By osmosis with the international traded sectors, wages in the private services, working for the local market, have to go up too. As their productivity is not going up at the same rate, cost-push inflation is created.

As productivity will increase

- hidden employment will disappear
- unskilled labor will have fewer opportunities.

Agriculture will have a hard time, even with the European subsidies, due to the low productivity (see Chapter 1, Table 1A3.1 of this Part II of the book).

Accordingly, unemployment will increase. The total labor force will increase as more young people will join the labor force and more women will participate, whereas labor will also emigrate. Anyway, it will be difficult to diminish the unemployment level under 12%.

Military expenses will go up. A full-professional army replaces the semi-professional army with recruits.

Progressive rates are installed for direct taxation, on the average bringing the figure at least from 33 to 40% of personal income. Real estate will be taxed.

The nuclear power plant is shut down.

Making the balance of all these structural changes it is unclear how the government budget will react. Contrary to the actual situation it is safer to assume a serious government deficit.

Finally, sovereignty is partly given up.

If some aspects have a rather pessimistic touch, one may not forget that perhaps the main advantages will be shown after 2011.

Scenario II

Lithuania in the Period 2002 – 2011 becomes a Full-Member of the European Monetary Union besides its Membership in the European Union, and in NATO (World or European wars excluded)

If the stabilization pact is maintained, the following economic indicators have to be respected:

Inflation	≤ 2 %
Budget deficit	≤ 3 % of GDP
Public debt	≤ 60 % of GDP
Balance of payments deficit	≤ 3 % of GDP

The situation of scenario one is maintained.

The level of the public debt is not an immediate problem. Given the description of scenario one, huge problems may arise for inflation, and the balance of payments.

Much depends on the way of sanctioning of the European Union. Will even forced secession be applied?

A tough time is perhaps the building stone for a brilliant long term future inside Europe.

Scenario III

Lithuania in the Period 2002-2011 is Not Yet a Full Member of the European Union, the European Monetary Union and NATO

(possibilities of a World or a European war are excluded)

One or more conditions of the table of Scenario II are not fulfilled.

All points of Scenario I are again realized though in a much slower way or are weakened in their realization.

Scenario IV

Full Membership in the European Union, the European Monetary Union and in NATO is given up by Lithuania in the Period 2002-2011 (possibilities of a World or a European war are excluded)

Instead of full membership in the European Union and in NATO, a loose cooperation is foreseen, namely under the form of a Free Trade Zone.

Stronger economic ties are forged with the other Baltic Sea Regions such as Belarus, Kaliningrad and the other parts of the Russian Federation, Latvia, Estonia, Poland, Norway, Denmark, Sweden, Finland and certain northern Länder of Germany, such as Schleswig-Holstein, Hamburg, Mecklenburg-Vorpommern and Berlin.

Some of these countries, such as Russia, Belarus, Latvia, Estonia, Norway and perhaps Poland may also only keep a loose cooperation with the European Union and also, with the exception of Norway, with NATO. Denmark and Sweden are not yet members of the European Monetary Union and Sweden not of NATO. Between the Länder of Germany on the one hand and Lithuania on the other, it is possible that the European Union allows regional contracts for economic cooperation. Economic growth will follow the cyclical movements in the mentioned states.

A status quo, more or less comparable with the fluctuations in the year 2001, is maintained concerning inflation, budget deficit, public debt, balance of payments, privatization, wages, productivity, employment and unemployment. This conservatism will satisfy large parts of the population, which is mostly afraid of changes.

The atomic plant is kept in operation and will continue to export electricity to the neighboring countries.

CHAPTER 3 - ANNEX 3

SCENARIOS FOR THE FUTURE OF THE FACILITIES SECTOR OF LITHUANIA 2003 - 2012

Based on the Nominal Group Technique, described in Annex 3 of Chapter 1 in this Part of the book, the following scenarios were deduced for the facilities sector of Lithuania for the period 2003-2012 (possibilities of a World or a European war are excluded).

Three scenarios are foreseen: the Fifth Resource Scenario, the Scenario of the Status Quo and the Cut Throat Competition Scenario.

The Fifth Resource Scenario

Foreign direct investments will come to Lithuania. They may find industrial zones with ready premises. However, they want more and they like to decide themselves on location. This is the moment that the fifth resource beside the resources of people, technology, information and capital, comes fully alive, namely the management of the Corporate Real Estate Assets. New companies will respond to this demand. These new companies of real estate assets management will look after space (location), design of buildings, construction, reparations, maintenance, waste management and eventually demolition. They will look after the direct investments so to say "from the cradle until the grave". This has to happen in the most effective way, i.e., with an Optimal Multiple Objective Utility. In this way, the foreign firms can look for an optimal Multiple Objective Utility for their People, Technology, Information and Capital, but also for their Corporate Real Assets.

The Status Quo Scenario

In this scenario, the situation in the facilities sector does not change. A set of small firms will remain operative in the facilities sector of Lithuania. By competition, aging, disagreement, etc. some small firms will disappear, but new ones will take their place.

The Cut Throat Competitive Scenario

In the Cut Throat Competitive Scenario the consequences of the productivity effect of the European Union will fully play. The productivity in the new member countries of the European Union will rise in the internationally traded sectors. The result is an increase in wages. The more national services have to raise their wages too, without an increase in productivity of the same size. This increase in wages will have an inflation effect in the country.

The Facilities Sector in Lithuania will fully undergo that influence. If it increases its prices, together with the other not internationally traded services, inflation will go up in Lithuania and ipso facto the cost of living will rise. Instead of increase in prices, diminution of quality of its services forms another alternative for the facilities sector. At that moment cut throat competition between the facilities management companies will occur. In this struggle for life, only the fittest will survive. It is also the moment that mala fide companies will appear, which will exploit the customers as much as possible.

NOTES PART 2 CHAPTER 3

[1] J.Foster, *Scenario Planning for Small Businesses*, Long Range Planning, Pergamon, 26, 1,1993, 125.

[2] K.Van Der Heijden, *Scenarios, the Art of Strategic Conversation*, Wiley, Chichester, 1996, 187.

[3] G.Ringland, *Scenario Planning, Managing for the Future*, Wiley, Chichester, 1998, 181.

[4] G.A.Miller, The magical number seven plus or minus two: some limits on our capacity for processing information, Psychological Review, 63, March 1965, 81-97.

[5] W.K.Brauers, Prévisions économiques à l'aide de la méthode entrées-sorties, Economica, Paris, 1995, 76.

[6] The combination of input-output and scenario writing is not new; see e.g. W.K.Brauers, *op.cit.* .76.

[7] These scenarios were presented at the Arusha Conference, hosted by the United Republic of Tanzania, September 12-18, 1999.

PART THREE

NEW ECONOMICS FOR A STAKEHOLDER SOCIETY

CHAPTER 1

INTRODUCTION TO MULTIPLE OBJECTIVE UTILITY THEORY

Some speak of Multiple Objective Utility Theory (MOUT), others about Multiple Attribute Utility Theory (MAUT). What is the difference? People during their entirely lifetime are moved by numerous objectives, which are considered here as independent from one another. A hierarchy of objectives exists going down from a foggy notion to measurable items, called attributes, which are then optimized. Objectives may be different for the several stakeholders: the public, the consumers, the firms, the government, whereas all kinds of pressure groups try to influence the choice of the objectives. Some speak of the theory of choice[1]. However, risk and uncertainty influence the choice, whereas probabilities and planning also enter the picture.

1.1 WHY FROM OLD TO NEW ECONOMICS?

A Belgian minister once launched the concept of *Gross Individual Happiness*, without knowing what this implies. Many elements will compose this happiness. Let us say that the individual is striving for several objectives and if they are fulfilled in an optimal way, happiness has to be reached. In other words, there is a problem of optimization of several objectives simultaneously. In addition, it will be difficult to satisfy every objective and to satisfy every individual. Therefore, we say with Steuer:

> "Since multiple objective problems rarely have points that simultaneously maximize all of the objectives, we are typically in a situation of trying to maximize each objective to the greatest extent possible"[2].

The Green Parties in Europe maintain that economic theory cannot solve this problem of human happiness. The reason is that they consider economics only as a science for minimizing costs. Therefore, we promulgate the idea of new economics, whereby all objectives are taken into consideration in an optimal way as much as possible. Of course, there is a problem of quantification, either ordinal or cardinal. Whereas insurance companies translate human life into monetary costs, here the value of human life is expressed in units related to e.g., maximization of health care or minimization of accidents.

A consumer buying a car faces several objectives. Indeed his objectives of safety, comfort, anti-pollution, and petrol consumption are in fact separate from the price of the car.

For the government the employment goal has to be an objective separated from its budgetary cost. This cost could include a compensation for unemployment. The unemployed, however, may not be satisfied with a monetary compensation alone. The government may also ask for pollution abatement and for new investments, for instance in public enterprises. Indeed most of these public enterprises show a shortage in investment.

An industrial firm aims on the one hand at an increase in productivity through a better motivation of personnel and through an improvement in the safety of workers and on the other at an increase in its market share. Additionally more and more a firm will also be pressed to satisfy ecological conditions "to the greatest extent possible". In the distribution sector for instance, one could strive for maximization of Net Present Value, minimization of the break-even period in the case of a new firm and to an optimal penetration in the European Union or in world markets. Project analysis may aim on the one hand at maximization of NPV, on the other at maximization of employment. Concerning privatization of public enterprises or services, the potential buyer will look for a minimum takeover price, a possible payback period, Net Present Value and Internal Rate of Return, whereas the government will look for a maximum takeover bid, employment, value added and new investments. Classical economics do not bring an answer concerning this multitude of objectives.

In fixing the objectives, one has to take into account all kinds of pressure groups, such as trade unions, employer organizations, consumer organizations and groups of ecologists or lobbies of all kinds. In addition, the state considers itself as the promoter of the general interest, ending up in Social Welfare or even in Social Well-being. All these agents may pursue different, even contradictory, objectives.

In order to define an objective better we have to focus on the notion of *Attribute*. Keeney and Raiffa present the example of the objective "reduce sulfur dioxide emissions" to be measured by the attribute "tons of sulfur dioxide emitted per year"[3]. An attribute should always be measurable[4].

Simultaneously we aim to satisfy multiple objectives, whereas several alternatives are possible, characterized by several attributes. A potential alternative should be quantitatively well defined. An attribute is a common characteristic of each alternative such as its economic, social, cultural or ecological significance, whereas an objective consists in the optimization (maximization or minimization) of an attribute. We avoid using the term "*Criteria*". Indeed, this term creates much confusion as being used for objectives as well as for attributes[5].

We focus on *General Aggregation* for multiple objectives[6]. In general aggregation for multiple objectives, the definition of Steuer is fully maintained. In addition, all objectives-attributes and all alternatives, which are stipulated from the beginning, are kept in the running. In the meantime, they are only dropped by filtering, given certain constraints, or by domination.

Partial Aggregation for multiple objectives considers general aggregation as unrealistic[7]. On the other hand, partial aggregation could remain stuck in incomparability and in order to come to a result, may use trial and error methods or a bit of juggling. Anyway, partial aggregation does not solve the problem of the simultaneous and optimal fulfillment of all objectives for a certain domain of the stakeholder society.

The theory to be developed here can be called *Multiple Objective Utility Theory (MOUT)* or *Multiple Attribute Utility Theory (MAUT)*.

1.2 VALUE OR UTILITY? THE RISK PROBLEM

No distinction has to be made between value and utility. Indeed, already in 1871, Jevons stated: "value depends entirely upon utility"[8]. Much earlier however, namely in 1853, Gossen remarked that:

> "the magnitude of value of a thing is measured exactly by the magnitude of life pleasure that it gives us"[9].

Value may be considered cardinal as well as ordinal[10]; important statement for our further research.

Nowadays, value is sometimes distinguished from utility. The utility function would be a value function including risk. We, however, return to the original definition. Once value and utility functions are considered identical, risk becomes a separate attribute of multiobjective utility theory, or a part of an attribute. In the last case, a risk factor could be linked to each attribute. Consequently, as several alternatives are considered, a *Risk Aversion Matrix* is obtained[11].

The following example illustrates this point. Suppose a fighter plane is still in the stage of research and development. A government co-financing the project will as an attribute include a risk factor between zero and one[12].

What is the relation between risk and *Uncertainty*? For Hertz and Thomas risk means both uncertainty and the result of uncertainty[13].

Fisher defines risk as follows[14]:

> "A risky situation is one in which the outcome is subject to an uncontrollable random event stemming from a known probability distribution, ... An uncertain situation, on the other hand, is characterized by the fact that probability distribution of the

uncontrollable random event is unknown".

Most authors consider uncertainty as a passive element dictated by the environment. Risk is more active The verb "to risk" means to expose to risk. Consequently, Loomba[15] identifies three types of attitudes: (1) a desire for risk, (2) indifference to risk and (3) risk aversion. Arrow states correctly that risk is subjective because risk bearing is not equal among economic agents[16].

Hertz, one of the pioneers of risk analysis, regarded it as: "a natural and logical extension of the sensitivity analysis approach"[17]. Decision analysts consider risk analysis as a special case of the more general decision analysis. Risk analysis however differs from decision analysis in its use of risk simulation "as a solution technique"[18].

Is Complete Uncertainty possible? In this context, one could try to move as soon as possible away from an *Uncertain Universe* to a more favorable universe. Let us take the example of buying by well-organized large companies. For them very often selling prices cannot be manipulated due to a severe market competition, while costs, being quasi-fixed, become overheads. The only possibility is then to face advantageous buying conditions, certainly for large firms such as steel works and large department stores, and the more so for firms which have to buy different raw materials and agricultural products in very large quantities. Indeed raw materials and agricultural products undergo heavily fluctuating world prices, which means that decisions are taken under pressure. Moreover, information, coming in by Internet or by telex, is sometimes incomplete, which means that buyers are not aware of the type and the size of the risk they have to take. For different products characterized by several attributes and facing several sellers, the important buying companies have to be assisted by a computer on which they can simulate several ratios fixed according to their own experience. This is a possibility of action in the case of quasi-complete uncertainty. In this sense, Arrow[19] is right when he maintains that situations, which are "basically certain", become uncertain for the individual, because the costs of transmission are not negligible. They vary from the type of information transmitted and from the education and the limited channel capacities of the recipients and the "coding" weakness of the sender.

In the case of portfolio analysis, an investor can limit the risk. Investments in corporate shares belonging to an uncertain universe (e.g., highly speculative) are counterbalanced by hedging or are replaced by investments in less speculative shares such as those of nationalized or semi-nationalized electricity producers or in corporate or official bonds.

Instead of an Uncertain Universe, one may try to move to other universes perhaps more certain, such as deterministic, antagonistic and probabilistic universes.

1.3 MULTIPLE OBJECTIVE UTILITY THEORY INSIDE THE THEORY OF THE STATISTICAL UNIVERSE

What is meant by the theory of the Universe? Here, Universe possesses a statistical, mathematical significance. The universe forms the set of all events. It is a set of sets, including for instance uncertain, deterministic, antagonistic, possibility and probabilistic universes. A completely Uncertain Universe does, in fact, not exist and a *Deterministic Universe* is not real.

The Deterministic Universe

In real life, a deterministic universe will usually boil down to a *Quasi-Certain Universe*. This was the case with the collectivistic and deterministic economic system of the former Soviet Union. The execution of the imperative planning was only quasi-certain due to climatological conditions, with for instance bad harvest years, human behavior and external influences, like worldwide depressions or recessions and international wars. Also, the growth economies of the 1960s in Western Europe and the United States showed quasi-certainty.

The Antagonistic Universe

Why does one undergo the "secret man" experience, i.e., the decision maker is purposely hiding his real motivation (not showing his cards?). The secret man acts in an Antagonistic Universe. Williams remarks that, if you think of a course of action, an opponent may foresee your plans and try to set a trap, which may capitalize on the choice you make[20] The problem is well known to card and chess players.

The Possibility Universe

In a possibility universe, the *Possibility-ratios* do not add up to 1. An example of Cross-Impact will make this clear, as is shown in Table 1.1.

Table 1.1 Possibility-ratios from Cross-Impact

Alternatives	Desired attribute I	Desired attribute II	Desired attribute III
Alternative A	$P_o = 0.6$	$P_o = 0.7$	$P_o = 0.6$
Alternative B	$P_o = 0.5$	$P_o = 0.5$	$P_o = 0.6$

For instance, the desired attribute II has a possibility ratio of 0.7, if alternative A is realized. This does not mean that attribute II is not reached for 30% or is reached for 30%, if alternative A does not take place. On the contrary, if alternative B is realized and not alternative A, the chance that attribute II is reached is equal to 50%. For instance, alternative A means the discovery of an extremely rich oilfield, resulting in a 0.7 chance of a significant decrease in oil prices. This does not mean that, if A does not take place, oil prices will not go down.

The Probabilistic Universe

In a probabilistic universe, the *Probability-ratios* add up to 1. In Table 1.1 the probability that the desired attribute II is not reached is 30%, if alternative A occurs. The probabilities of several actions under several events, found for instance after several Delphi rounds, are represented in decision trees.

In terms of the usual *Decision Tree Analysis*, the difference between risk and uncertainty is clearly underlined. Indeed, circles indicate uncertainty, under the form of chance events, preferably described in scenarios. Boxes indicate risk taken under the form of a decision.

Suppose that for sales of commodities a choice is possible between two actions (a box). For each action, three scenarios are possible: success, slowing down or failure. Each scenario counts two subscenarios, namely with or without competition. Probabilities are estimated for each of these situations. If gains or losses are estimated for each scenario or subscenario, the total of the net gains will show which action is the best.

If uncertainty and risk are based on observation parameters, another form of variability may appear. Indeed, *Random Coefficients* reflect the fact that the parameters are not the same for every observation of an event. At that moment, the random problem can be considered in two ways viz., by standard deviations or by distribution functions.

The size of the *Standard Deviation* measures the spread or the variability of the event[21].

Thomas and Hertz suggest the use of statistical distribution theory after a normal distribution with calculation of the mean and variance of the probability distribution of the magnitude of the attributes[22]. From expert experience, a frequency distribution is obtained for each of these attributes per alternative. What is the form of the frequency distribution in this connection e.g., in project analysis for cash flows and Net Present Value (NPV)? In the case of non-normally distributed cash flow variables, is their total normally distributed? Under certain assumptions the Net Present Value (NPV) distribution is approximately normal, if the number of years of the project's life is sufficiently large. Bonini, experimenting with Monte-Carlo simulations, concludes that distributions of NPV are usually skewed[23]. Hull reports that the distribution of NPV is approximately normal in three out of five case studies[24]. Thomas, in an earlier study, found normality in the NPV distribution[25]. All these findings mean that firm statements about normality of NPV cannot be made.

Anyway, it is clear that probabilities and random coefficients can be present together in one or another application.

What conclusions are drawn for multiobjective theory? We simply propose to speak of an *Expected Multiobjective Utility Theory*, whereby risk factors are incorporated as an attribute of the alternatives or/and as a part of the alternatives.

1.4 MEASUREMENT OF ATTRIBUTES BY INDICATORS

Usually attributes are measured by pieces, in money terms or in Si-units[26].

What about the effectiveness of an army, the effectiveness of a school system, the effectiveness of management, the counterpart of the indirect costs and the overheads of a firm? Therefore, more and more firms and governments are working with indicators, norms or ratios. In fact, all the stakeholders would have to do it. The consumer unions are doing it increasingly concerning the satisfaction or dissatisfaction of the consumers.

Anyway, indicators or norms have to fit in a system and these systems were introduced in the firms and in the government.

Planning, Programming, Budgeting System (PPBS) originated in the private sector but became mainly operational in the American Defense Department and afterwards in the American public sector. Other names cover in fact the same meaning as PPBS. President Nixon coined the term *Program Evaluation*[27]. In Canada, the term *Evaluation Analysis* is officially used. In France, the official name is *RCB (la Rationalisation des Choix Budgétaires)*. The older term *Program Budgeting* is also still used.

Even in managerial economics, confusion is general for systems similar to these of the public sector such as *Corporate Planning, Corporate Strategy, Management by Objectives (MBO)* and *Organization Development (OD)*.

In order to avoid this general confusion of terms, the common characteristics of all these public and private sector systems are summarized here as *Prospective Planning System (PPS)* (see Fig. 1.1).

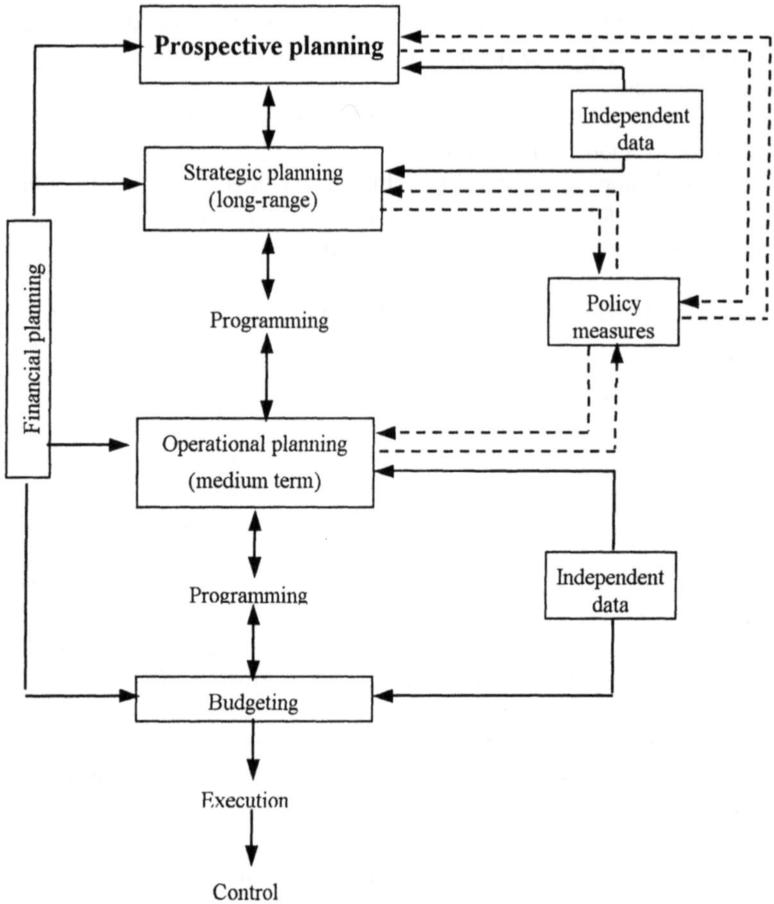

Figure 1.1 The Prospective Planning System (PPS)

Prospective Planning

The prospective plan is a plan which fixes the objectives structure of the institution under consideration. This structure covers long term spans. The prospective plan will define and evaluate the alternatives to achieve the objectives selected. However, policy makers cannot change some independent data[28]. Some policy measures form fixed data for certain institutions. Policy measures imposed by the government form given data for a firm, though certain firms may influence the policy measures of the government. The prospective plan is much interested in research and development.

Strategic Planning

The strategic plan is a long range plan, which is mainly concerned with growth investment and equipment, population and personnel, but also with the supply of raw materials (long term contracts). The period of the strategic plan depends on the characteristics of the activity. The defense planning of the U. S. normally uses an eight-year strategic plan. A large European steel consortium is based on a strategic plan for 8 to 12 years, but in some subsectors of the chemical industry, such as the photographic and pharmaceutical industries, the strategic plan is limited to a five-year period.

Operational Planning

The operational plan forms medium term planning with the inclusion of social costs, whether or not expressed in money terms, and the use of averages (e. g., in defense, the consideration of an abstract, average escort vessel).

The operational plan deals with the connection between all inputs, production and distribution, and is involved with investment, mainly replacement investment.

Budgeting

Budgeting is planning for the short term, mostly a year. The budget is mainly directed towards current revenue and expenditure.

Programming

Programming bridges the differences between strategic and operational plans and between the operational plan and budgeting. Let us consider the latter case. The program translates the operational plan into reality. The program for several years rotates, since every year the most recent year is dropped and a new year is introduced. This means that every year a new program is born, which has also to be consistent with the prospective planning.

The budget covers the first year of the program. This budget at least has to be expressed in real items. In the case of the American Navy for example, the budget considers each ship separately, either new or worn out. For government services the budget is rigid. Intermediate changes can only be made by Congress. This manner of acting contrasts with the habits of the private sector. In the firm, the budget is unchangeable. Expenditure, however, may deviate from the budget on condition that both are compared afterwards. For the revision of programs and plans, formal rules are set up for government services. Private firms will act in the same way. Essential to the system is that costs should be linked to outputs.

Financial Planning

The basis of financial planning is formed by costs on condition that these costs are real costs for the institution concerned and are expressed in monetary terms. The financial planning covers the whole period from the budgeting period to the very long term of the prospective planning. In this sense, financial planning is also a bridge plan.

In financial planning all costs over this long period are discounted to their present value in order to make comparisons possible. Moreover, in some institutions important equipment and material may have a life spanning several planning periods. This is the case with defense when acquiring, for instance, an assault plane to be financed over ten years and with an expected life of fifteen years. The financing of such important equipment and material is inflexible due to long-term contracts. In this way, there will be some antagonism between the various funds necessary in the different planning periods. The prospective planning tries to acquire a maximum of research and development funds. The strategic plan is mainly interested in important expansion investment; the operational planning mainly in replacement investment; the budgeting in current expenditure. The financial planning has to harmonize all these tendencies as a function of the objectives.

Within the financial planning, the capital model will describe the acquisition of the funds needed and the characteristics of these funds. For a firm these characteristics will be: internal financing, external and even international financing, co-operation, merger and government participation.

In the case of government services it is important to know how the financing of expenditure will operate over such a very long period. Generally, governments do not carry out such a financial planning.

Zero Base Budgeting (ZBB)

After Phyrr, Zero Base Budgeting was developed as the latest cure for financial problems or as a tool for squeezing maximum output from a controlled budget[29].

At Rand Cooperation David Novick, the pioneer of PPBS, said to the author: Zero Base Budgeting is a "fraud". If this is slightly exaggerated, it is true that ZBB means a "tabula rasa" for the past, with setting up of other decision packages. These decision packages are reviewed and ranked all the time, certainly each budget year, after not very clearly defined objectives.

Being a complete break with the past, a general application is not advisable. In 1977, the newly elected president Jimmy Carter brought with him to Washington the idea of Zero Base Budgeting to improve the operation of the federal government departments. This was not very successful. A limited application is rather advisable, unless the environment is completely changed.

At the occasion of the end of the Cold War, the reorganization of military defense in the United States presented a good example of a possible general application of ZBB. Lacking a global enemy after the end of the Cold War, the United States moved to a two-major-regional-conflict strategy (nearly simultaneously in e.g., Korea and the Persian Gulf)[30]. This strategy was recently changed to a one-major-conflict strategy.

New political targets for military action, such as "Peace Keeping" and "Peace Enforcing", could be considered as new fields for applying ZBB in the Western World.

1.5 MEASUREMENT OF QUALITY

Notwithstanding what was said earlier concerning the introduction of indicators, norms or ratios, measurement may remain difficult for some qualitative aspects. Some imagination is necessary, eventually by looking for an *Alternative Measurement*. For instance in residential construction, the attribute, price per square meter of the terrain in question, measures "district prestige". On the other side, aesthetics in residential construction remains as a problem. At that moment, a Nominal Scale is sometimes used.

A *Nominal Scale* gives only labels or names as appreciation and nothing more, such as: good, high, important. The information is very fuzzy, unless a certain ranking is already included, such as: moderate, good or very good. Even then, the information remains fuzzy, unless each of the labels receives a number.

The comparison of scales could be for instance:

10 very low
20 low
50 moderate, but satisfactory
80 high
100 very high[31].

Such comparison is chosen in a very arbitrary way. Therefore, certain

systematization is necessary, such as with an Arithmetical or a Geometrical Progression.
Sometimes an *Arithmetical Progression* is used under the form:

0 (non-existing), 1, 2, 3, 4, 5, 6, 7, 8, 9, 10

A nominal scale is linked to this numerical scale. The jump from more or less low to low is 1 and that from not high to high is also 1, whereas the difference is much more significant for the last case.

Another Arithmetical Progression could be:

10 very low
20 low
30 moderate, but satisfactory
40 high
50 very high

Here also the jump from low to moderate and from moderate to high is only 10, although, the difference is more significant, as compared with very low to low.
Therefore, some categories are added:

10 very low
20 low
30 more or less low
40 moderate, but satisfactory
50 more or less high
60 high
70 very high

The choice of categories remains arbitrary.
Saaty makes another proposition inside his pair-wise comparisons: the *Fundamental Scale*[32].

1	equal importance
3	moderate importance of one over another
5	essential or strong importance
7	very strong importance
9	extreme importance

When compromise is needed, 2,4,6,8 are intermediate values between the two adjacent judgments.

A *Geometrical Progression,* for instance under the form 1,2,4,8,16... presents a solution for solving the problem of the distances between the categories.

Some authors go even further. Lootsma promotes the *Normal Scale*[33]:

$$e^0 = 1;\ e^1 = 2.7;\ e^2 = 7.4;\ e^3 = 20.1\ \ldots$$

This scale seems exaggerated, which is even more the case with his *Stretched Scale*:

$$e^0 = 1;\ e^2 = 7.4;\ e^4 = 54.6;\ e^6 = 403.4\ldots$$

The mentioned geometrical progression is a current phenomenon in nature. Already Malthus indicated the geometrical progression in the increase of the human population. In biology, mitosis is a cell division, which leads to geometrical progression in the number of cells.

Anyway, the comparisons between *Nominal Scales* and *Dimensionless Numbers* seem disputable. It is better that from the beginning as much as possible nominal scales are avoided.

Measurement of quality has to be defined, neither by a decision-maker-dictator, nor by researchers, but by the stakeholders, i.e., people affected by the issue. They are asked to estimate an attribute of an objective by the choice of a dimensionless *Cardinal Number*. This may happen in several ways, such as by using questionnaires or by telephone interviews.

In the case of aesthetics in residential construction, the task of people interested in the issue becomes easier, if they can compare several kinds of residential constructions at the same time. They are asked to rank their aesthetical perception in a limited range. A scale from 1 to 7 is considered as a maximum[34]. In case of several stakeholders the danger exists that some preferably will choose in the higher range of 1 to 7 and others in the lower range. Therefore, a more limited choice is preferred, for instance, from 1 to 4. This range may be still too large as 2 is the double of 1 and four equals four times one. Perhaps the scale 2, 3, 4 or the scale 2, 3, 4, 5 are preferable. Zero can still be added to the scale for somebody who disapproves completely.

As a conclusion, first we prefer direct measurement, or at least alternative measurement. If this is not possible, persons interested in the issue will be asked to show their understanding by giving a dimensionless cardinal number to the attribute of an objective.

1.6 HIERARCHY OF OBJECTIVES. WHAT DOES HIERARCHY MEAN?

Two kinds of hierarchy can be considered: on the one hand a complete a priori priority between the objectives and on the other hand a hierarchy under the form of a pyramid descending from a *Top Objective* to *Subobjectives* and to *Sub-subobjectives*.

1) A complete a priori priority

The hierarchy in a complete a priori priority starts from the assumption that a complete a priori priority exists between the objectives, which is dictated by the stakeholders beforehand. If quantification is possible, consecutive linear programming will achieve the result. The approach will be extensively discussed later. In fact, an a priori priority boils down to a mono-optimization by stages.

This explanation concerns alternatives automatically generated during the process, i.e., in the presence of a continuous set of alternatives. In the discrete case, i.e., with a limited series of possible alternatives, the stakeholders also rank the objectives after their importance. This approach, which is also called "*the Lexicographic Method*", is mostly considered as too simplistic[35]

2) A pyramid of objectives

In the second kind of hierarchy a *Hierarchical Objectives Structure* exists under the form of a pyramid with *Subobjectives* and *Sub-subobjectives* descending with increasing specificity and decreasing permanency[36]. The specificity finally boils down to measurability. It is a top-down approach contrary to the bottom-up approach of Roy[37].

If General Well-being is the top objective Economic Welfare, Individual Well-being and Ecological Balance could be the subobjectives. At that level of objectives measurability is still absent but at the next lower level measurement is possible by *Indicators* such as response time, income and employment for Economic Welfare; public goods, life expectancy and security for Individual Well-being and water, air and noise pollution for Ecological Balance.

Is it advisable to count as much as possible an equal number of these sub-subobjectives per subobjective? This balancing equilibrium is not necessary, as all sub-subobjectives have to be present in each alternative. However, it is advisable to boil down as much as possible a very large number of sub-subobjectives to a smaller one. In addition, in the pyramid of objectives the *Substitutive Principle*, giving power to a lower versus a higher authority with a priority for the lower one, may play a role.

1.7 A HIERARCHY OF OBJECTIVES FOR THE DEMOCRATIC REPUBLIC OF CONGO

With the pyramid of objectives a hierarchical objectives structure exists under the form of a pyramid descending to measurability. Applied to the Democratic Republic of Congo, General Well-being would be the top objective with Economic Welfare, Individual Well-being and Ecological Balance as the subobjectives. For practical reasons, the sub-subobjectives are limited to Health Care, Security and

Education concerning Individual Well-being; to Employment and Basic Needs concerning Economic Welfare; to Air Pollution concerning Ecological Balance. At this level measurability is possible (see Figure 1.2)[38].

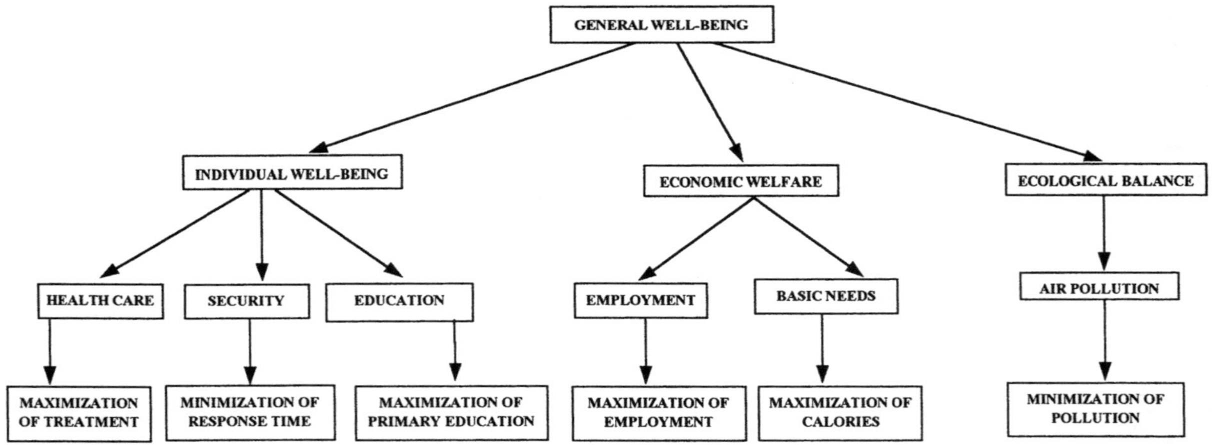

Fig. 1.2 The pyramid of objectives as applied for the Democratic Republic of Congo

The existing structures are temporarily maintained. Here an evaluation technique has to be used comparing the outcome of several structures. For education, for instance, the existing schools are maintained, though evaluated in comparison to one another, whereas an additional effort is foreseen in the pyramid of objectives. Considering the existing structures the Planning, Programming, Budgeting System (PPBS) is applied, whereas in other cases a start from tabula rasa is preferred with application of Zero Base Budgeting (ZBB). It is the case for Security and Pollution.

If a number of projects is proposed in order to fulfill the objectives, they have to stay inside a new fixed budget, which is provided outside the conventional budget. The use of the conventional budget is evaluated inside PPBS, as was described

An example of application is given in the following Figure 1.3.

PROJECTS	extra Health Care	ZBB in Security	extra Education	extra Employment	extra Basic Needs	ZBB for Anti-pollution
Project I	Total patients treated per year with preventive medicine	Centralized but heavily mechanized police force in the cities	Primary education with tv.-monitors and built in microphones (videos, CD) only seconded by teachers	Possible employment over 5 years in new industrial zones	Free distribution of basic nutrition	Severe control on air pollution from factories in the cities
Project II	Total patients treated per year with curative medicine	Decentralized labor intensive police force in the cities	Primary education mainly based on teachers	Possible employment due to subsidies for first installation	Irrigation policy	Severe control on air pollution from traffic in the cities
Project III	Total patients treated per year with preventive medicine	Centralized but heavily mechanized police force in the cities	Primary education mainly based on teachers	Possible employment due to subsidies for first installation	Free distribution of seeds, fertilizers and phyto-sanitary products	Severe control on air pollution from factories in the cities
Project IV	Total patients treated per year with curative medicine	Centralized but heavily mechanized police force in the cities	Primary education with tv.-monitors and built in microphones (videos, CD) only seconded by teachers	Possible employment due to subsidies for first installation	Irrigation policy	Severe control on air pollution from traffic in the cities
Project V	Total patients treated per year with preventive medicine	Decentralized labor intensive police force in the cities	Primary education mainly based on teachers	Possible employment over 5 years in new industrial zones	Irrigation policy	Severe control on air pollution from traffic in the cities
Project VI	Total patients treated per year with curative medicine	Decentralized labor intensive police force in the cities	Primary education with tv.-monitors and built in microphones (videos, CD) only seconded by teachers	Possible employment due to subsidies for first installation	Free distribution of basic nutrition	Severe control on air pollution from factories in the cities
Attributes	Totals of patients per year: three preventive treatments equal one curative treatment	Response time in minutes	Increase in % of children in that age category over 5 years (target UNO 80%)	Total of possible employment increase over 5 years	% increase of calories for total population over 5 years	% CO_2 diminution in the cities per person over one year
Objectives	MAXIMUM	MINIMUM	MAXIMUM	MAXIMUM	MAXIMUM	MINIMUM

Fig. 1.3 Projects with a fixed budget and responding to the pyramid of objectives for the Democratic Republic of Congo

1.8 CONFLICT RESOLUTION

Conflicts between stakeholders are considered as conflicts between objectives. Some stakeholders may support a green environment and pollution abatement. Others would like to promote economic growth. Multiple optimization will try to optimize each objective as much as possible. Hierarchy between stakeholders is still another possibility[39]. The same approach is followed as for hierarchy with objectives.

However, there are cases where convergence does not work. With ideologies, fanaticism, prejudices etc. it is possible that any logical reasoning is absent. At that moment, multiple optimization may not function. Other methods like simulation analysis, especially gaming, may help (see Part II, chapter 2).

1.9 INSTRUMENTS FOR A NEW ECONOMIC POLICY

The comments, which were formulated in Part I mostly concerned methods of *Forecasting*, whereas the purpose of this book is mainly to give instruments for a new policy. The distinction is not so sharp. Forecasting may lead to a *Self-fulfilling Prophecy*. In addition, a forecast in demography over fifty years may show the aging of the population and may lead the government to revise its policy on pensions.

Some objections may arise that the same methods of optimization with multiple objectives are proposed for the private and for the public sector. This is not a relevant point. Both private and public sector have their own independent objectives, though increasingly they are forced to some similar objectives in what is called *Sustainable Development*. More and more private firms e.g., will have to participate in pollution abatement and will have to guaranty the security of their workers and of their products.

Value or Utility? For the classical authors they are identical, but for contemporary authors utility function is a value function including risk. We however support the idea that risk is an attribute or/and a part of the alternatives.

Multiple Objectives Utility considers an optimal utility for all objectives simultaneously. For Multiple Attribute Utility, utility is included in the measurability of the attributes themselves, as in Si-units, in monetary terms or in indicators, norms or ratios. Indicators, norms or ratios are defined by a decision-maker-dictator, or by the government, considering itself as the promoter of the general interest (in systems such as PPBS, Zero Base Budgeting) or inside a firm (MBO, OD). From our point of view, all the stakeholders concerned have to fix the indicators, norms or ratios. Purely qualitative measurements have to be avoided as much as possible.

The next chapters will start with a demonstration of several concepts of multiobjective utility theory and namely with the traditional ones such as: cost-benefit, cost-effectiveness, the weight concept and the concept based on a series of binary comparisons. All these methods have to be critiqued on the following grounds:

- total optimization versus mono-optimization would mean optimization of all objectives "to the greatest extent possible". Mono-optimization, being the case of optimization for one objective, is also the case if objectives mostly weighted are added up to a single super-objective, which is then optimized;
- the economic law of decreasing marginal utility has to be respected.

In a filtering process, one could stop alternatives not respecting certain constraints. In addition, one could look if an alternative does not dominate the other alternatives for all objectives, a problem-solving outcome. If filtering and domination do not help, incomparable alternatives will remain. This remark will form the core of the theory, which will be presented here.

It will be demonstrated that a utility function for multiple objectives has to be:

1) nonadditive with no weights (with a chance for an in-between alternative);
2) nonlinear, as a result from indifference curves analysis. With the assistance of indifference curve analysis, we should end up with several other concepts.

Indifference curves, however, are difficult to determine in practice, but Reference Point Theory can be used as an approximation, with normalization by scores. Indifference Approach, Multiplicative Method and Ratio Analysis are still other nonlinear methods, which are worthwhile to be considered. These methods will avoid the cautionary comments which were cited above.

NOTES PART 3 CHAPTER 1

[1] S.H.Heap et al. *The Theory of Choice, A Critical guide*, Blackwell, Oxford, 1992.

[2] R.E. Steuer, *Multiple Criteria Optimization: Theory, Computation and Application*, R. E. Krieger Publishing Cy. Malabar, Florida, 1989, 138.

[3] R.L.Keeney, and H.Raiffa, *Decisions with Multiple Objectives. Preferences and Value Tradeoffs*, Cambridge University Press, USA,1993, 32.

[4] R.L. Keeney, and H. Raiffa, *op. cit.38.*

[5] C-L. Hwang, A. S. Masud, *Multiple Objective Decision Making; Methods and Applications,* Springer, Lecture Notes in Mathematical Systems, Berlin, 1979, 6-7.

[6] The term "General Aggregation" is chosen in comparison with the term "General Equilibrium Models" (see Part I, 2.3 of this book). In addition, the term "General Economic Equilibrium" has been used in economics from the time of Quesnay on (see J.Balderston, *Models of General Economic Equilibrium*, in: Economic Activity Analysis, O.Morgenstern (ed.) John Wiley, New York, 1954.

In aggregation literature the term "General Aggregation" is in general use. (J.Balderston and T.Whitin, *Aggregation in the Input-Output Model*, in: Economic Activity Analysis, op.cit, 79-128.

[7] In the same way, economic theory speaks of a partial approach (see Part I,1.3 of this book).

The term "partial aggregation" for multiple objectives comes from A.Schärlig, *Décider sur plusieurs critères*, Presses polytechniques romandes, Lausanne, 1985, 139-205.

[8] W.S.Jevons, *The Theory of Political Economy*, 4th ed., Macmillan, London, 1911, 1.

[9] N.Georgescu-Roegen, *H .H. Gossen: his Life and Work,* in: Historical Perspective, Introductory Essay to: *The Laws of Human Relations*, H. H. Gossen translated in English, MIT Press, Cambridge (Mass), 1983, XVI.

[10] W.J.Shaver, *On the existence and interpretation of value allocation*, Econometrica, vol.48, n°2, March, 1980, 467-476.

[11] G.T.Duncan, *A matrix measure of multivariate local risk aversion*, Econometrica, May 1977, 895-903.

In the same line of thought and in comparison with independent objectives, the notion of *Risk Independence* is introduced. A set of risk attributes is independent from the risk of any other objective.

R.L.Keeney, *Risk Independence and multiattributed Utility Functions*, Econometrica, vol.41, 1973, 27-43.

L.G.Epstein, *Multivariate Risk Independence and functional forms for Preferences and Technologies*, Econometrica, vol.48, May 1980, 973-986.

[12] Sometimes the risk factor is expressed in money terms: the *Risk Premium*. The risk premium is the reduction in wealth that an individual will accept to eliminate a random variation. The risk premium is very well known in portfolio analysis. If risk aversion is high, the risk premium is high: R.E.Kihlstrom, D.Romer, S.Williams, *Risk Aversion with Random Initial Wealth*, Econometrica, vol.49/4, 1981, 911-920.

[13] D.B.Hertz, H .Thomas, *Risk Analysis and its Applications*, J. Wiley, Chichester, 1983, 3.

[14] G.H.Fisher, *Cost Considerations in Systems Analysis*, American Elsevier, New York, 1971, 202.

[15] N.P.Loomba, *Management - A Quantitative Perspective,* Collier Macmillan, New York, 1978, 30.

[16] K.J.Arrow, *Political and Economic Evaluation of Social Effects and Externalities*, in: M.D. Intriligator, Frontiers of Quantitative Economics, North-Holland/American Elsevier, Amsterdam - New York, 1971, 13.

[17] D.B.Hertz, H.Thomas, *op. cit*, 1983.,19.

[18] D.B.Hertz, H.Thomas, *op . cit.*, 1983, 21.

[19] K.J.Arrow, *op. cit.*, 1971, 14-15.

[20] J.D.Williams, *The Complete Strategist*, The Rand Series, McGraw-Hill, New York, 1954, 13.

[21] Ahituv and Spector give special attention to variance in multiobjective theory. N.Ahituv, Y, Spector, *Evaluating Information under two Criteria: Expected Payoff and Risk*, Working Paper Nr 70/90, I.I.B.R., Tel Aviv University, Israel, 1990.

[22] D.B.Hertz, H.Thomas, *op. cit.*, 1983, 41-42.

[23] C.P.Bonini, *Risk Evaluation of Investment Projects*, Omega, 3 (6), 1975, 735-750.

[24] J.C.Hull, *The Input to and Output from Risk Simulation Models*, European Journal of Operational Research, 1, 1977, 368-375.

[25] H.Thomas, *The Assessment of Project Worth with Applications to Research and Development*, in: J.N.Wolfe, Cost-Benefit and Cost-Effectiveness, Allen and Unwin, London, 1973, 88-117.

[26] Si-units have a broader meaning than units of the metric system do. The metric system goes back to the time of Napoleon with as basis the meter and the kilogram. Si-units come later (Système International d'Unités, 1960) and are taken over in the legislation of the several countries (cf. ISO-International Standard Association, European Normalization Institute, American National Standard Institute). Meter-kilogram-second forms the basis as a translation of power, energy, mass and acceleration.

[27] Weekly compilation of Presidential Documents, week ending May 30, 1970, Washington D.C.

[28] "Among the data there are some that can be changed - to a greater or lesser degree - by policy-makers. They will be called 'means', ... There remains the class of means which, as a rule, are of a quantitative character and are used for frequent changes, in fact for the adaptation of the economy to small and frequent changes in some of the other data. This class will be called 'instruments' or 'instrument variables' ". J.Tinbergen, *Economic Policy: Principles and Design*, North-Holland, Amsterdam 1956, 4-5.

[29] P.A.Phyrr, *Zero Base Budgeting, a practical management tool for evaluating expenses*. John Wiley, New York, 1983.

[30] Z.Khalilzad, *Strategic Appraisal 1996*, Project Air Force, RAND, 1996, 1-8.

[31] In our mind, these dimensionless numerals are *Cardinal Numbers*. An *Ordinal Scale* is given by: *1st, 2nd, 3rd, 4th*, ...

[32] T.L.Saaty, *What is the Analytic Hierarchy Process?* in: Mathematical Models for Decision Support, Nato Advanced Study Institute, Val d'Isere (F), July 26,1987, Mon 82

[33] F.A.Lootsma, *Numerical Scaling of Human Judgement in Pairwise-comparison Methods for Fuzzy Multi-criteria Decision Analysis*, in: NATO Advanced Study Institute, op. cit. Thu 39.

[34] G.A.Miller, *The magical number seven plus or minus two: some limits on our capacity for processing information*, Psychological Review, 63, March 1965, 81-97.

[35] H.Voogd, *Multicriteria evaluation for urban and regional planning*, Pion limited, 1983,45.

[36] W.K.Brauers, *Systems Analysis, Planning and Decision Models*, Elsevier, Amsterdam, New York, 1976, 34-37.

[37] B.Roy, *Méthodologie multicritère d'aide à la décision*, Economica, Paris, 1985, 173 and foll.

[38] This section is a part of a report written for one of the governments of president Kabila Sr.

[39] G.Anandalingam, V.Apprey, *Multi-level programming and conflict resolution*, European Journal of Operational Research, vol. 51 N° 2, 1991, 233-247.

CHAPTER 2

SOME TRADITIONAL METHODS TO APPROACH THE PROBLEM OF MULTIPLE OBJECTIVES

Some traditional methods were thought to hold solutions for the multi-objective problem. Cost-Benefit e.g., uses one single unit, namely money. In that way, it represents an extremely materialistic approach. The person who is eager to work will be very unhappy with a purely monetary compensation. A child who lost his parents in a car accident is not compensated for his grief with a monetary payment. Security on the roads is not a purely monetary problem, but also a problem of education and training, of soberness and road and weather conditions.

Cost-Effectiveness is already a better method, if there was not the dual problem concerning cost and effectiveness.

The Scorecard Method brings a lot of interesting information, but a clear-cut proposal for the stakeholders is not reached. The weighting concept produces only extreme solutions. After the weighting concept an alternative, which is always good but never excellent in an objective, will have no chance to be selected. Binary comparisons were already condemned in the eighteen-century.

2.1 COST-BENEFIT ANALYSIS AND DECISION TREE ANALYSIS

Cost-Benefit Analysis is a method with a monetary unit as the common unit of measurement of benefits and costs. Indeed even benefits are expressed in the chosen monetary unit, either in a direct or in an indirect way. Ipso facto the net-benefit, is either positive or negative. The proposed solution is then, either the acceptance of the project, or the status quo ante, which sometimes go hand in hand with deteriorating circumstances.

In traditional cost-benefit analysis, only one single project is considered. Eventually it would be possible to take into account several projects in order to compare their respective net-benefits with the original situation.

Anyhow, cost-benefit presents a materialistic approach, whereby for instance unemployment and health care are degraded to monetary items.

People are more easily solution-minded rather than objective-oriented. Cost-benefit analysis is a product of this way of thinking. For instance, cost-benefit about a new underground railway in London starts with thinking of an eventual construction of that railway[1] and not of objectives such as slimming of London, diminishing of transport flows: home-work, home-supplies, home-entertainment and office-office. At that moment, entirely other solutions interfere, such as teleworking, teleshopping and teleconferences.

Cost-benefit of the road system of a country[2] and cost-benefit of the inland river transport system of a country[3], even for one new canal[4], were typical solution minded cost-benefit studies. For transport of passengers, collective transport systems have to be considered too and for freight: inland river and railway transportation versus road transport. As the world is getting more and more objective-minded, such as thinking of ecology, cost-benefit studies will have fewer and fewer chances today than before.

Decision Tree Analysis is in a certain way comparable to cost-benefit, as most of the time only money costs are involved. Anyway, decision tree analysis is mono-objective, but in addition, probabilities are involved. In 1994, the dilemma was raised in China, either to update the old industry or to replace the old industry by new investments, either in the same sector or in other sectors[5]. The choice was proposed between a new tool industry and the modernization of the existing tool industry represented by a decision point (a rectangular) in the decision tree. For each choice the following chance events (represented by a circle in the decision tree) could happen on the market: the product succeeds as estimated, sales slow down or the product is a complete failure. Each time competition or no competition is also foreseen. Probabilities were estimated for each case. The result over fifteen years was more than positive for the new industry.

There is, however, not only the point of view of the firm, but also that of the government with its income from taxes. In addition, macroeconomics is interested in discounted Value Added and in the influence on the balance of payments. Finally the Internal Rate of Return, the Payback Period of entrepreneurial economics and the influence on employment are not possible to express in money terms. Ipso facto they were not included.

2.2 COST-EFFECTIVENESS ANALYSIS AND FRACTIONAL PROGRAMMING

In Cost-Effectiveness, several projects are taken into account simultaneously. The analysis, however, is bi-objective: costs expressed in a common monetary unit on the one side and a single effectiveness-indicator on the other. For instance in defense a weapon system could balance costs against the rate of kill, expressed in one or another military indicator[6].

As initially optimality was absent in cost-effectiveness, several addenda were proposed.

First, Lange launched his Economic Principle. The *Economic Principle of Lange* runs like this: either costs are kept constant with maximization of an objective (*Effectiveness*), or effectiveness is kept constant with minimization of costs (*Efficiency*)[7]. From linear programming it is known that for this dual the solution is identical, which is only an assumption for nonlinear systems. At that moment, the question remains if an optimal solution is found.

Second, Fractional Programming forms a substitute for the dual problem:

$$\max.E / C = \max.\ \text{effectiveness}/\min.\text{costs}$$

For privatization the fractional programming formula is reversed:

$$\max.\ \frac{C}{E} = \frac{\max.\ \text{takeover price at the government side}}{\min.\ \text{unemployment}}$$

Other attributes of objectives for privatization are, however, excluded such as: return on equity and investment commitments for the takeover party, the change in the general tax level, the expected value added, the influence on the balance of payments and all other financial, monetary, regional and ecological repercussions. Indeed, in Fractional Programming, not all objectives are included[8]. Moreover, it is not so clear that fractional programming produces the optimal solution in the Steuer sense[9].

2.3 THE SCORECARD METHOD

In 1953 the islands of the province of Zeeland in the Netherlands were flooded causing the death of thousands and thousands of persons and billions of Guilders in material losses. Closing the islands with one huge dam, making the island a part of the continent and changing the Ooster-Schelde estuary in a huge sweet water reservoir, presented a good solution for the safety of people, but was found very harmful by the ecologists. Higher dikes on the islands were accepted by the ecologists but not by the safety people. In this way the Ooster-Schelde problem was a good example of conflicting interests defended by several groups.

The breakthrough would come by finding a solution acceptable for all the parties and helpful to their objectives. It means an effort of creative thinking. Brainstorming may be useful in this context, but also the Scorecard method. This method was invented by the Rand Corporation in the case of the Ooster-Schelde[10]. All the advantages and disadvantages of the two propositions were detailed in a systematic way with tables, graphs and figures. On the basis of this information, new solutions were looked for, bringing a kind of greatest possible divisor for all opinions. The solution found consisted of storm dams in front of the islands, which would weaken the floods but keep the Ooster-Schelde as an open estuary with salt water. This solution was satisfactory to both security officers and ecologists, and the proposition finally passed in the Dutch Parliament. Perhaps another objective was overlooked at that time, namely the minimization of the budget costs, given the increasing cost price of these gigantic public works.

It is clear that it will not always be possible to find a compromise solution.

2.4 THE WEIGHTING CONCEPT

We follow Schlaifer's definition for weighting[11], namely that weights are defined as the importance:

> "which a decision maker attaches to each of the events in a set of mutually exclusive and collectively exhaustive events".

Schlaifer attaches the following rules to this definition:

> "Rule 1. The sum of the weights assigned to any set of mutually exclusive and collectively exhaustive events shall be 1".
>
> "Rule 2. The weight assigned to any event shall be a number between 0 and 1 inclusive".
>
> "Rule 3. If two or more mutually exclusive events are grouped into a single event, the weight attached to the single event shall be equal to the sum of the weights attached to the original events".

These rules stress the additive property of weights.

Cost-benefit is a narrower interpretation of the weighting method, whereby this time the weights are money units. Measuring employment and unemployment in money units is, however, a very materialistic approach.

The Discrete Case

In the discrete case, a number of well-defined and possible alternatives is known[12]. The decomposable utility function over *n* attributes of the objectives

$$f[u_1(x_1), u_2(x_2), \ldots, u_i(x_i), \ldots, u_n(x_n)]$$

is the best known under its additive form with weights:

$$x_j = \sum_{i=1}^{i=n} w_i \, x_{ij}$$

with:

$$\sum_{i=1}^{i=n} w_i = 1$$

x_j = possible alternative j

$j = 1,2,...,m$; m number of alternatives
x_{ij} = attribute i of alternative j
$i = 1,2,...,n$; n number of attributes
w_i = weight for attribute i

This formula represents a *Convex Linear Combination* of the attributes[13].

If for an objective the minimum of an attribute is required, this attribute is provided with a negative sign in the additive form.
A ranking of the x_j alternatives produces the final result
By using weights, however, a single *Superobjective* replaces the several multiple objectives.

Example

Several alternatives generate jobs and value added.

Objective 1: max. of man-years $w_1 = 0.4$

Objective 2: max. of. VA $w_2 = 0.6$

One superobjective (mono-objective) replaces the two objectives, namely:

Maximization of a unit package composed of:
40% man-years and
60% value added.

The Continuous Case of a Generating Process

A continuous set of alternatives, with alternatives generated during the process, makes the use of linear programming possible for optimization:

$\boldsymbol{a'x} = max.$
with as constraints:

$\boldsymbol{Bx \leq c}$
with: $\boldsymbol{x}$ as a vector of attributes
$\boldsymbol{a}$ as a vector of weights
$\boldsymbol{B}$ as a matrix of coefficients
c as a vector of constraints

If for an objective the minimum of an attribute is required, the minimum is reversed to a maximum by a fraction.

Example (see Figure 2.1)
For a new firm the break-even years (i.e., the number of years needed to realize the first profit) have to be minimized and the Net Present Value has to be maximized under the constraints of two production units with limited capacity. The two production units work as one group.

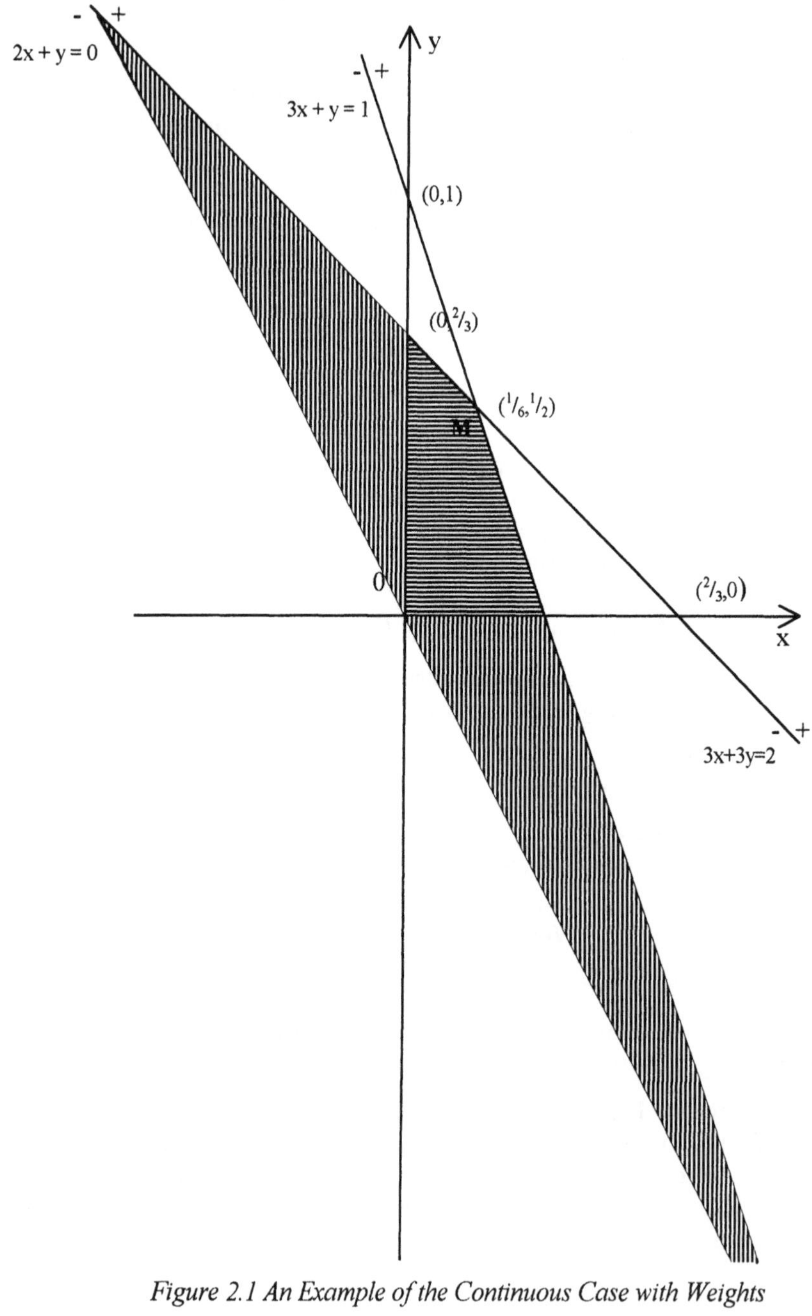

Figure 2.1 An Example of the Continuous Case with Weights

Assuming: x: the number of break-even years reversed with weight 2/3
y: NPV in million $ with weight 1/3

results in the objective function: $2x + y = max$.

With constraints:

for production unit one: $3x + 3y \leq 2$
for production unit two: $3x + y \leq 1$
with: $x \geq 0$
$y \geq 0$

Solution

$2x + y = 0$ is determined by the points: $x = 0 \;\; y = 0$
$x = 1 \;\; y = -2$

Common point for the constraints, which is reachable for the objective function:

$$\begin{array}{rl} 3x + 3y &= 2 \\ \underline{-3x - y} &\underline{= -1} \\ 2y &= 1 \end{array} \rightarrow \underline{y = 1/2} \qquad \begin{array}{rl} 3x + 3y &= 2 \\ \underline{-9x - 3y} &\underline{= -3} \\ -6x &= -1 \end{array} \rightarrow \underline{x = 1/6}$$

with: M(1/6,1/2)

Conclusion

NPV = 1/2 million dollars
Break-even point after six years

The constraints could lead to several maxima. In the example, if a constraint is formed by a line with equation: $2x + y = a$ with: $a > 0$, then an infinite number of solutions, i.e. maxima, exists.

In the example of the firm, the formula of convex linear combination can be used under the assumption of constant returns to scale with a stable production function. Sooner or later, however, decreasing marginal utility will interfere at the demand side.

The Bilinear Case

Weights are just as feasible in a bilinear multiobjective approach as in the semi-multiplicative representation of Keeney and Raiffa[14]; e.g., for two attributes:

$$u(y,z) = k_y u_y(y) + k_z u_z(z) + k_{yz} u_y(y) u_z(z)$$
$$\text{with: } k_y + k_z + k_{yz} = 1$$

Here also by using weights the attributes are added together, which means that a single *Superobjective* is created that is to be optimized.

2.5 THE APPROACH BASED ON A SERIES OF BINARY COMPARISONS

The Condorcet Paradox

In 1785 Condorcet was not satisfied with the generally used form of voting. Starting from that view he examined the possibility of giving votes to different rankings of candidates, whereas the final decision was taken after the *Plurality Rule* i.e., in favor of the candidate receiving the greatest number of votes for all its first places[15]. His several binary combinations led to an unexpected *Condorcet winner*[16] and to what was generally called the *Condorcet Paradox*[17] after he demonstrated the nonsense of binary comparisons for multifarious conclusions[18].

We use a Condorcet example after replacing the different rankings of candidates by different rankings of alternatives per objective, whereas scores per objective replace the numbers of votes.

The algorithm finally consists of a multifarious generalization of binary comparisons between possible alternatives facing several objectives. This algorithm could lead to a contradictory outcome as illustrated by the example.

Alternatives A and B receive scores of the stakeholders for the realization of five objectives:

objectives		1	2	3	4	5
scores		23	17	2	10	8
ranking	1st	A	B	B	A	B
	2nd	B	A	A	B	A

A P B (P means preferred to) with A scoring 33 against a preference for B of 27.

If however three alternatives are present, the outcome could be contradictory.

Indeed:

objectives		1	2	3	4	5
scores		23	17	2	10	8
ranking	1st	A	B	B	C	C
	2nd	B	C	A	A	B
	3rd	C	A	C	B	A

A P B P C with A scoring 23, B 19 and C 18.

Binary comparisons would have produced:

> between A and B: A (with 33) P B (with 27)
> between C and A: C (with 35) P A (with 25)
> between B and C: B (with 42) P C (with 18).

This is Contradictory indeed.

For Bordes and Tideman[19] *Arrow's General Possibility Theorem*[20], which is rather an Impossibility Theorem, is mainly based on the manipulation of candidacies in voting, for us alternatives. The result could be a reverse order. Let us give an example for consistency conditions concerning two alternatives A and B:

objectives		1	2	3
scores		9	7	4
ranking	1st	A	B	B
	2nd	B	A	A

If the plurality rule is used then: B (with 11) P A (with 9). Suppose a third alternative is introduced, viz., C, in such a manner that:

objectives		1	2	3
scores		9	7	4
ranking	1st	A	B	C
	2nd	B	A	B
	3rd	C	C	A

leading to: A (with 9) P B (with 7) P C (with 4) i.e., a *Reverse Order* compared to the binary case. In this way the introduction of C could be a maneuver. Indeed Bordes and Tideman remark that for public works projects with two alternatives the legislative branch may favor B. If the executive branch prefers rather A, it will introduce a new project C, for which it does not care, but which will insure that A is chosen above B.

Dropping the plurality rule, contradiction or reverse order are also possible if points are attributed to the alternatives after their ranking e.g., two points for a first ranking, one point for a second ranking and zero points for an eventual third ranking. Let us illustrate that again with the Condorcet example.

objectives		1	2	3	4	5
scores		23	17	2	10	8
ranking	1st	A=46	B=34	B=4	A=20	B=16
	2nd	B=23	A=17	A=2	B=10	A=8

A P B with A scoring 93 against a preference for B of 87.
If however three alternatives were present the outcome could be different. Indeed:

objectives		1	2	3	4	5
scores		23	17	2	10	8
ranking	1st	A=46	B=34	B=4	C=20	C=16
	2nd	B=23	C=17	A=2	A=10	B=8
	3rd	C=0	A=0	C=0	B=0	A=0

leading to: B (with 69) P A (with 58) P C (with 53) i.e., a reverse order compared to the binary case.

For a contradictory outcome the soccer example is misleading. Soccer team A beating soccer team B, soccer team B beating soccer team C and then C beating A, is indeed misleading as C is beating A at a later moment when A and C have changed their characteristics (other team composition, other strategy, other training, other physical and psychological attitudes, other weather conditions and supporters assistance etc.). This misleading result is mostly weakened by playing at least two matches, one on the home field of C and one on the home field of A.

Other Illustrative Examples

Drink consumption may lead to contradictory results too as e.g., wine is preferred to milk and milk to beer, whereas beer is preferred to wine. Introducing several objectives such as satisfaction of thirstiness, enjoyment and nutrition capacity and allotting ten points to each objective, the outcomes may be of a reverse order or contradictory.

Drink consumption	thirstiness	enjoyment	nutrition	total
Wine	2	8	4	14
Milk	8	2	6	16

i.e. milk P wine

Drink consumption	thirstiness	enjoyment	nutrition	total
Milk	0	2	6	8
Beer	10	8	4	22

i.e. beer P milk

Drink consumption	thirstiness	enjoyment	nutrition	total
Wine	0	8	6	14
Beer	10	2	4	16

i.e. beer P wine

General conclusion of binary comparisons: beer P milk P wine.

Making a multifarious comparison may give:

Drink consumption	thirstiness	enjoyment	nutrition	total
Wine	0	6	3	9
Milk	2	0	4	6
Beer	8	4	3	15

i.e. beer P wine P milk: a *Reversed Order*.

Separate binary comparisons may also lead to a contradiction:

Drink consumption	thirstiness	enjoyment	nutrition	total
Wine	2	8	4	14
Milk	8	2	6	16

i.e. milk P wine

Drink consumption	thirstiness	enjoyment	nutrition	total
Milk	0	2	6	8
Beer	10	8	4	22

i.e. beer P milk

Drink consumption	thirstiness	enjoyment	nutrition	total
Wine	2	8	6	16
Beer	8	2	4	14

i.e. wine P beer

This means that: **beer P milk P wine**, which is in contradiction with: **wine P beer**.

Considering all these cases, it is clear that generalization of binary comparisons in facing multiple objectives have to be avoided. Therefore, multifarious comparisons of alternatives are preferred instead of binary ones.

NOTES PART 3 CHAPTER 2

[1] C.D.Foster, M.E.Beesley, *Estimating the social benefit of constructing an underground railway in London*, Journal of the Royal Statistical Society, 1963, 46-92.

[2] G.Blauwens et al. *Kosten-batenanalyse van de verdere uitbouw van het Belgisch Wegennet,* Department of Public Works, Brussels, 1982.

[3] M.Anselin et al. *Kosten-batenanalyse van de verdere uitbouw van het Belgisch Waterwegennet,* Department of Public Works, Brussels, 1982.

[4] W.Nonneman, *Investeringsanalyse van een duwvaartverbinding Oelegem-Antwerpen*, Bond beter Leefmilieu, Brussels,1975.

[5] W.K.Brauers, *Scenarios for Updating or Replacement of Old Industry*, High Technology and its Development Strategies, Wuhan (China), 1994, ISBN 7-5375-1341-4/TB.3, 519-533.

[6] W.K.Brauers, *Systems Analysis, Planning and Decision Models*, Elsevier, Amsterdam - New York, 1976, 67-126.

[7] O.Lange, *On the Functioning of the Socialist Economy*, Szkola Glowna Planowania i Statystyki, Warsaw,1968, 66-70.

[8] In econometrics a single *Disturbance Term* includes all the remaining objectives. P.Kennedy, *A Guide to Econometrics*, Blackwell, Oxford, 1998, 2.

[9] In fact, we considered the discrete case in fractional programming. Nykowski and Zolkiewski are interested in the continuous case. See: I.Nykowski, Z.Zolkiewski, *On some connections between bicriteria and fractional programming problems*, Essays and Surveys on Multiple Criteria Decision Making, Ed. P. Hansen, Springer, Berlin, 1983, 300-309.

[10] Rand Corporation, *Protecting an Estuary from Floods - A Policy Analysis of the Oosterchelde,* Prepared for the Netherlands Rijkswaterstaat, Santa Monica (Cal), December 1977.

[11] R.Schlaifer, *Probability and Statistics for Business Decisions*, Mc Graw Hill, New York, 1959, 8-11.

[12] More details on the distinction between discrete and continuous choices are given in: K.A.Small and H.S.Rosen, *Applied Welfare Economics with Discrete Choice Models*, Econometrica, vol.49, n°1,1981, 105-130.

[13] R.G.D.Allen, *Mathematical Economics*, Macmillan, London, 1957, 384-387.

[14] R.L.Keeney, H.Raiffa, *Decisions with Multiple Objectives. Preferences and Value Tradeoffs*, Cambridge University Press, USA, 1993, 234.

[15] Marquis de Condorcet, *Essai sur l'application de l'analyse à la probabilité des décisions rendues à la pluralité des voix*, l'Imprimerie Royale, Paris, 1785, LVIII-LXI.

[16] R.Kast, *La théorie de la décision*, édit. La Découverte, Paris, 1993, 62.

[17] R.Kast, *op.cit.*,63.

[18] In the 1963-edition of his book Arrow maintains that in the first edition of 1951 he was not aware of the work of Condorcet:

> "when I first studied the problem and developed the contradictions in the majority rule system, I was sure that this was no original discovery, although I had no explicit reference, and sought to express this knowledge by referring to the well known 'paradox of voting"

K.J.Arrow, *Social Choice and Individual Values*, (2nd edit.) Yale University. Press, New Haven, 1963, 93 (First ed. Wiley, New York, 1951).

[19] G.Bordes, N.Tideman *Independence of irrelevant alternatives in the theory of voting*, Theory and Decision, 1991,182-183.

[20] K.J.Arrow, *op.cit.*,1963, 13,15,97.

CHAPTER 3

MULTIFARIOUS COMPARISONS FOR OBJECTIVES AND ATTRIBUTES

The previous chapter concluded that multifarious comparisons of alternatives are preferred above binary ones. For multifarious comparison, this chapter moves from some theoretically self-evident approaches to situations in which the alternatives seem to be incomparable. Indeed, an a priori priority between the objectives gives the impression that the problem is easily solved. This seems also the case, if some alternatives are dropped out, due to not satisfying some minimum or maximum requirements. In addition, domination of one alternative over all the others would easily solve the problem. The problem remains if a ranking turns out to be impossible with incomparably looking alternatives.

3.1 THE APPROACH OF A COMPLETE A PRIORI PRIORITY BETWEEN THE MULTIFARIOUSLY COMPARED OBJECTIVES

The approach of a complete a priori priority between the multifariously compared objectives starts from the assumption that this priority exists between the objectives. Beforehand, the stakeholders dictate this complete a priori priority. In fact, this priority boils down to a mono-optimization by stages.

Distinction has to be made between a discrete and a continuous case. The discrete case faces a set of a limited number of alternatives, whereas the continuous one concerns alternatives, generated out of a set of continuous and numerous alternatives.

The Discrete Case

In the discrete case, the objectives are ranked after the sequence, dictated by the stakeholders. Holmes proposed therefore the name of *Lexicographic Method*[1].

The alternative with the optimal (maximum or minimum) attribute for the first ranked objective is chosen, irrespective of the attributes of the lower ranked objectives.

Suppose for instance that the policy makers of a country decide upon the following a priori sequence:

1) growth rate of Gross Domestic Product (Value Added)
2) employment
3) investments
4) imports.

If the growth rate of policy A is 3%, of policy B is 2% and of policy C is 2.5%, then policy A is chosen irrespective of the level of employment which is generated through A. If however the growth rate of policies A and C is 3%, then the employment level is considered. Suppose policy A creates 250,000 jobs and policy C 300,000 then policy C is chosen etc.

Some will call *Lexicographic Method,* the method where the priority levels mean "much greater than". For instance, GDP counts for 50%, employment for 30%, investments for 15% and imports for 5%. Once again, the weighting process creates a *Superobjective* (see for instance[2]).

The Generating Process in the Continuous Case

Consecutive linear programming will present the final results[3]. We return to the input-output model (see Part I,2.4). Traditionally the production structure of a country is presented by[4]:

$$(I-a)x$$

x the vector of the production per sector
y the vector of the final demand
$v'x$ value added (GDP)
$l'x$ labor requirements
$b'x$ capital requirements
$m'x$ importation levels

First optimization stage

The Value Added of the country obtains the first priority:

$$\text{Max. } v'x$$

subject to:

$$(I - a)x \leq y$$

$$l'x \leq L \qquad L = \text{full employment}$$

$$b'x \leq C \qquad C = \text{the existing full capacity in investment}$$

$$m'x \leq M$$

M = the import level not to be attained

Second optimization stage

Labor requirements obtain the second priority:

$$Max.\ l'x$$

subject to:

$$(I - a)x \leq y$$

$$l'x \leq L$$

$$b'x \leq C$$

$$m'x \leq M$$

$v'x$ = a constant: the optimal value added (GDP), as obtained in the first optimization stage

Third optimization stage

Capital requirements obtain the third priority:

$$Max.\ b'x$$

subject to:

$$(I - a)\ x \leq y$$

$$m'x \leq M$$

$v'x$ = a constant: the optimal value added (GDP)

$l'x$ = a constant: the optimal labor requirements

Fourth optimization stage

Importation levels obtain the last priority:

$$Min.\ m'x$$

subject to:

$$(I - a)x \leq y$$

$v'x$ = the optimal value added (GDP)

$l'x$ = the optimal labor requirements

$b'x$ = the optimal capital requirements

The final outcome will determine the production per sector x.

With a complete a priori priority of the objectives, the continuous case, more than the discrete one, takes into account the lower ranked objectives. A discrete case for the example would take, for instance, the maximization of value added into consideration exclusively and absolutely unless an ex aequo would exist; otherwise an alternative with a lesser value added but with a higher employment level would be excluded from the beginning.

In the continuous as well as in the discrete case, it has to concern strong, complete and not confused a priori priorities of the stakeholders. If two objectives are strongly linked to each other in an interdependent way, , such as turnover and prices, *Quadratic Programming* can be used in one of the stages.

If from the side of the stakeholders no a priori priority for the objectives exists, other possibilities have to be explored.

3.2 THE FILTERING STAGE

Some of the candidate alternatives are excluded if they do not respond to conditions concerning lower bounds or upper limits.

A "pater familias" likes to buy a new car. He has the choice between five cars: A, B, C, D and E. The objective "comfort" possesses the following attributes: excellent, medium and weak. The comfort of car E is weak. A weak comfort is in no way acceptable for the pater familias. Consequently, car E does not pass the filtering stage. The weak comfort forms a lower bound: the comfort has to be better than weak, namely medium or excellent.

Car D costs $75,000. The budget to spent by the pater familias is only $50,000 maximum. The amount of $50,000 is an upper limit and car D is considered no more.

a. Lower Bounds

$x_i \geq g_i$ with: $i = 1,.., n$; n being the number of attributes.
However, some of the lower bounds g_i may not exist.

b. Upper Limits

$x_i \leq f_i$ with: $i = 1,.., n$.
However, some of the upper limits f_i may not exist[5].

c. Hard and Soft Constraints

All constraints concerning lower bounds or upper limits have to be hard constraints, which form a sine qua non for the acceptance of the candidate alternatives[6].

Distinction has to be made between qualitative and quantitative hard constraints.

On the one hand, investments needed in a well-defined region and not in other regions, complete financial guaranties to be given for daughters of multinationals in case of failure, represent examples of qualitative hard constraints.

On the other side, certain capacities in production not to be exceeded unless new investments are made, the World Bank granting a loan to a developing country unless for instance at least an Internal Rate of Return of 12% is guaranteed, geometrical constraints under the form of a limiting line, surface or manifold, represent quantitative hard constraints examples.

By definition, soft constraints are not very clearly delimited. The payback period of an investment cannot be fixed as a hard constraint. Indeed a payback period of three years would for instance exclude in such a way an investment with a payback period of three years and one month. Of course, to determine a border line between a soft and a hard constraint may look difficult. To maintain the same employment level as before (e.g., 500 persons) could be a hard constraint, but the creation of 500 additional jobs is a soft constraint, otherwise an alternative only offering 499 new jobs but with many other advantages would be excluded. Therefore, this last soft constraint will be rather included in a next stage, the ranking stage. In the ranking stage, for the objective of job creation, the alternative of creating for instance 501 new jobs will be preferred above the alternative creating 499 jobs.

Several competing alternatives, or only one, could be taken into consideration. In the last case, the only one alternative has simply to pass the filtering stage[7]. Several competing alternatives, however, have to not only pass the filtering stage but also have to face the following stages.

3.3 THE POSSIBILITY OF A DOMINANCE STAGE

The concept of dominance will help to set up a ranking of candidate alternatives, once they have passed the filtering stage. Dominance plays a prominent part in multiobjective optimization.

Let us return to the pater familias buying a car. He has still the choice between cars A, B and C.

Car A shows the following attributes:

price: $45,000
comfort: excellent
speed: medium
shape: ordinary

Car B:

price: $40,000
comfort: excellent
speed: fast
shape: special

Car C:

price: $35,000
comfort: excellent
speed: fast
shape: special

Our pater familias will choose car C.

We say that car C dominates car B and that car B dominates A. In this case, there is also *Transitivity:* C also dominates A.

A distinction has to be made between strong and weak dominance in the several objectives of the candidate alternatives[8].

a. Strong Dominance

The vector of its possible attributes represents each alternative; for instance alternatives z and y by vectors z and y respectively.

The vector of attributes z strongly dominates the vector of attributes y, iff $z > y$ i.e., $z_i > y_i$ for all attributes i, with $i = 1, 2, ..., n$; n being the number of attributes; > denotes a strict order. Strong dominance decides about the order of more than two alternatives, for instance, z P y P x P u (P = Preferred).

If one alternative dominates strongly all the others for all attributes, then this alternative is ranked first. An alternative, dominated by the others for all attributes, is ranked last.

b. Weak or Ordinary Dominance

The vector of attributes z weakly dominates the vector of attributes y, iff $z_i \geq y_i$ for all i, and $z_i > y_i$ for at least one i and not for more than n-1 attributes with $i = 1, 2,...,n$ and n as the number of attributes.

A contradictory alternative ranking lower than an already chosen one with which it is in contradiction is excluded. An alternative, repeated several times, is given several rank positions and variations of an alternative are ranked separately[9].

3.4 DOMINANCE AND THE LAW OF DECREASING MARGINAL UTILITY

Gossen formulated the economic law of decreasing marginal utility for the first time[10]. Edgeworth-Pareto visualized this law under the form of indifference curves for two commodities[11].

Now, the whole theory of indifference curves is applied for multiobjective analysis, not for goods and services, as usual, but for objectives. Gossen already demonstrated that the law is valuable for immaterial goods and ideas[12]. Ipso facto, the law is valuable for utility considerations in attributes and objectives.

Illustrated for two objectives in multiobjective utility theory, very interesting conclusions are drawn for dominating and dominated alternatives and for efficient points (see Figure 3.1).

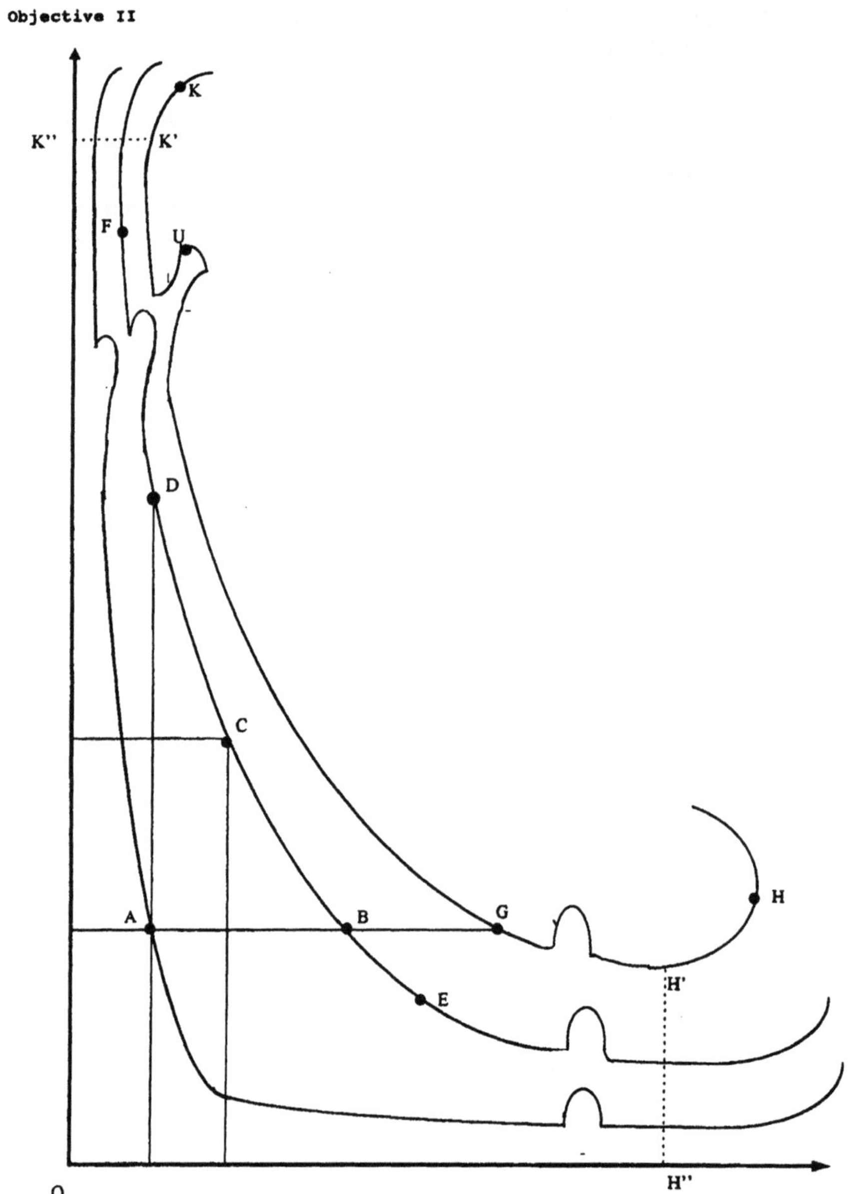

Fig.3.1 Indifference Curves and Multiobjective Utility Theory

At first sight, it is clear that points D, C, B and G dominate A. Following indifference curve theory, new dominating points are discovered. Additionally, points E and F dominate A. Consequently the multiobjective utility function for two objectives has to be represented by an indifference curve, possessing all the characteristics from indifference curve analysis. This leads to the discovery of new dominance effects. Indeed an alternative situated on an indifference curve is always dominated by an alternative on a higher indifference curve but alternatives situated on the same indifference curve automatically take the same ranking.

Fig.3.1 shows a multitude of indifference curves forming an indifference map. Which indifference curve does represent the multiobjective utility function? The highest one is selected, which is possible after the constraints on the available resources. The highest indifference curve KH represents the multiobjective utility function, if corresponding constraints on the available resources are given and if all points of KH pass the filtering stage. On the contrary, if for instance upper limits are enforced in the filtering stage on objective II (e.g., K'K") and on objective I (e.g., H'H"), then only a part of the indifference curve, namely K'H', satisfies all conditions. The final conclusion is extremely interesting: all dominating alternatives belong to the highest possible indifference curve and trade-off between these alternatives becomes possible.

Indifference curve analysis forms a perfect illustration of the economic law of decreasing marginal utility. Increasing marginal utility is also possible. Vickrey[13] accepts the possibility of a perverse slope at the ends of an indifference curve, which is comparable to the well-known perverse slope of the demand curve known as the Giffen Paradox[14]. Decreasing as well as increasing marginal utility stress the nonlinearity of indifference curves. Consequently, constant marginal utility has to be excluded.

In addition, the *Nonconvexity* of the indifference curves has to be stressed. *Convexity* or *Nonconcavity* follows here the Minkowski definition: a convex set has the characteristic that all points on a line between any two points of that set have to belong to that set[15]. However, a stakeholder could prefer a temporary increase in marginal utility for a specific objective at any moment (e.g., point U of Figure 3.1). This is acceptable as long as the indifference curves do not intersect. Indeed intersection would mean two different utilities for the point of intersection.

Indifference curve analysis influences the continuous case too. In Figure 2.1 of Chapter 2 of this Part of the book, the case was presented of a firm with a rigid production function, resulting in a maximum production at point M. However, this firm at the supply side has to take into account the demand side too. The demand side after decreasing marginal utility may result in another equilibrium point than M. In Figure 3.2 point M' represents the new maximum. Indeed, point M' is situated on a higher indifference curve than point M.

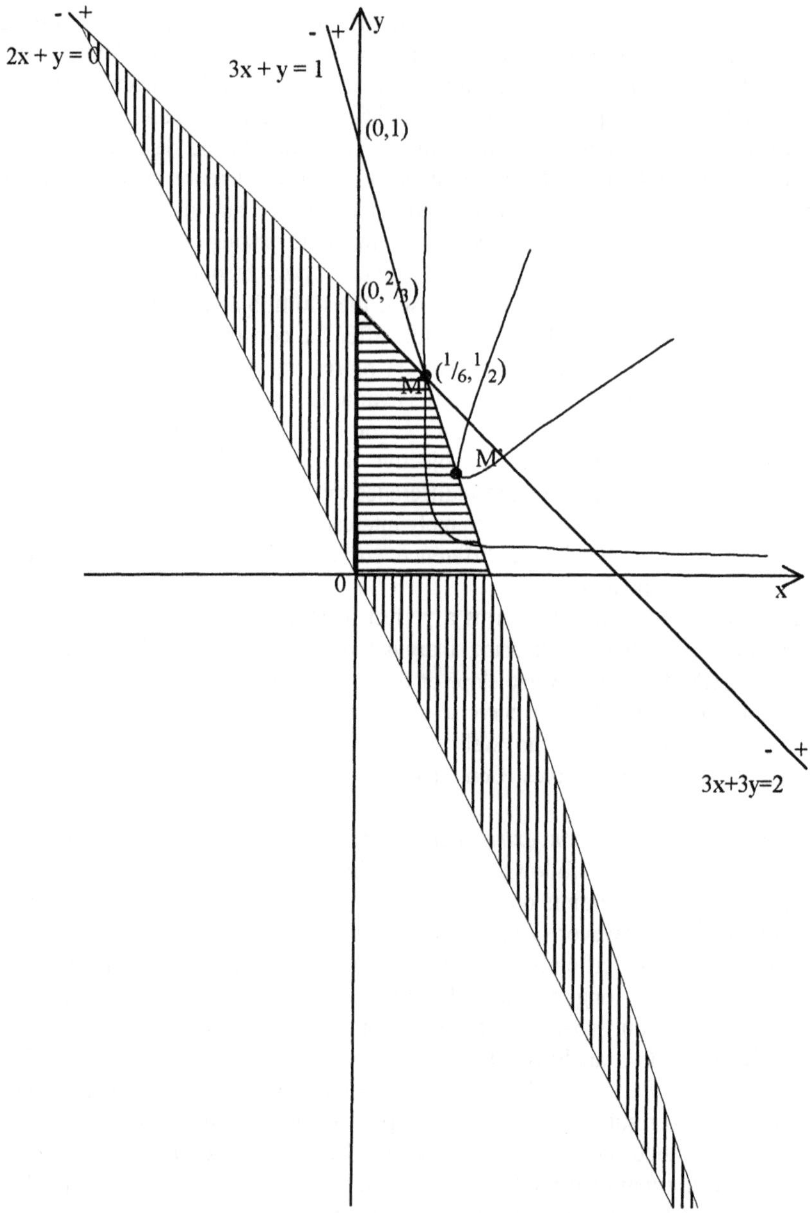

Figure 3.2 Continuous Case with the introduction of Indifference Curves

The firm under consideration produced under *Constant Returns to Scale*. However, possible *Increasing Returns to Scale* or *Decreasing Returns to Scale* will show, once again, *Nonlinear Curves*.

Over time, still other changes are possible. Point M' may change again, namely by trade[16]. Even more, the *Nonlinear Life Cycle* of a product may interfere. For instance, this cycle may evolve as follows. In the beginning, a slow growth pattern is shown, as the new product is not yet known, followed by an exponential growth if the product becomes successful. The saturation point is reached when for durable consumer goods only the demand for replacement remains. Finally, the product will disappear by changes in the needs or by substitution by other products. During the lifetime cycles of products, the shape of the indifference curves will change.

Once again, nonlinearity seems to dominate, with the exception of the temporary constant returns to scale in a firm.

Linearity being excluded, it could be suggested that *Piecewise Linearity* might be acceptable as a compromise, because any convex or non-convex function can be approximated by piecewise linear ones. Piecewise Linearity, however, leads to an additive form with weights, which means the creation of a *Superobjective,* replacing the multiple objectives.

Similar conclusions as for two objectives can be drawn for three, four or more objectives, with an *Indifference Surface* corresponding with three objectives and an *Indifference Manifold* with more than three objectives. In this way indifference curve analysis as described above presents a visualization of the nonlinear multiobjective utility functions.

Later on *Community Indifference Curves*, for instance in international trade, were proposed[17]. This means that one can go farther than individual behavior. For multiobjective utility theory community indifference curves signify that indifference curves are known for different stakeholders together.

At that moment, the trade-off between the alternatives on the highest possible indifference curve, as mentioned earlier, becomes a trade-off between the aspiration levels of several stakeholders. This trade-off was in the mind of Pareto when he wrote:

> "we say that the members of a group benefit of the maximum of utility at a certain position if it is impossible to move away a little bit from this position without increasing or decreasing the utility of each individual of that group. It means that even a small deviation from this position will necessarily increase the utility for certain individuals but decrease the utility for the others; to be agreeable to the first, disagreeable to the last"[18].

Is it possible to define the Pareto Optima ?

Pareto called a point an optimum *(Pareto Optimum)* when it is not possible to move away from that point without hurting one or another party. Consequently, if the *Indifference Loci* (indifference curve, surface or manifold) show the aspiration level of the stakeholders, each point belonging to the highest possible *Indifference Locus*, given limited resources, represents a Pareto optimum.

Such approach announces a new way of thinking. Mostly, it was accepted that for each set of data only one single Pareto Optimum existed. Now it is clear that this is not true. ***For a given set of data, several Pareto Optima are possible simultaneously. Pareto Optima pave the way for Multiobjective Optimization.***

Nevertheless, it is doubtful that all these Pareto optima are definable in reality. Indeed observations are mostly insufficient to determine in a continuous way all Pareto points of an indifferent locus. In the case of a limited number of discrete points, additionally imaginary Pareto optima could complete an indifference locus. Their position, however, is not certain. Consequently, a complete set of Pareto optima is unknown. At that moment, one has rather to think of substitutes for indifference curve analysis. These substitutes have to approach as much as possible the Pareto theory, namely by taking into consideration nonlinearity, Pareto-optima and even nonconvexity. Anyway, this means that methods based on linearity or on weights have to be omitted.

In this line of thought, a valuable proposition is labeled: the Indifference Method.

NOTES PART 3 CHAPTER 3

[1] J.C.Holmes, *An Ordinal Method of Evaluation*, Urban Studies, Vol.9,N°1,1972,179-191.

[2] D.Giokas, M.Vassiloglou, *A goal programming model for bank assets and liabilities management*, European Journal of Operational Research,vol.50, N°1, 1991, 48-60.

[3] R.E.Steuer, *Multiple Criteria Optimization: Theory, Computation and Application*, Krieger, Malabar (Fla., U.S.), 1989, 292, develops a similar approach and calls it "preemptive goal programming".

[4] The full details of the model with simulations not only for one country but also for the economic integration of several countries are given in: W.K.Brauers, *Input-Output Analysis and International Economic Integration* (in Dutch with English Summary), Standaard Uitg., Antwerpen-Utrecht, 1968. W.K.Brauers, *Prévisions économiques à l'aide de la méthode entrées-sorties,* Economica, Paris, 1995.

[5] We represented lower bounds and upper limits by scalars, but these constraints could also be linear functions. In fact in the example of Chapter 2.4 and in Figure 3.2 the constraints are linear functions. Even nonlinear constraints are acceptable. The underlying idea we developed for the linear constraints still apply for the nonlinear ones. For nonlinear constraints, see: L.E.Edlefsen, *The Comparative Statics of Hedonic Price Functions and other Nonlinear Constraints*, Econometrica, vol.49, 1981, 1501-1520.

[6] A.P.Wierzbicki, M.Makowski, *Multi-Objective Optimization in Negotiation Support*, WP-92-007, IASA Laxenburg (A), 1992., 4.

[7] As an example, see: W.K.Brauers, *Multiple Criteria Decision Making in Industrial Project Management*, Engineering Costs and Production Economics, Vol.20, 1990,.231-240.

[8] R.E.Steuer,*op.cit.*, 1989,147.

[9] Is domination absolute? Weitzman seems to deny it (M.L.Weitzman, *Optimal search for the Best Alternative*, Econometrica, vol.47,3, May 1979, 641-654). There could be a problem of search known as *Pandora's Box*. Weitzman shows an application of *Research and Development*. Two new technologies are considered: alpha and omega. Alpha dominates strongly omega. Alpha has lower research cost, shorter development lag, higher expected return, greater minimum return and less variance than omega. Nevertheless Weitzman proves that it is better to develop omega first.

The example is completely theoretical. The final return is only known when the products are on the market. The problem is not sequential: try omega first and by failure. try alpha. The problem was just to make a choice on the basis of expectations. The expectations for alpha were: a 0.5 probability of a return of 100 and 0.5 of 55. It is not excluded that the final return of alpha is, for example, 80, but probably never 240. The expectations of omega were of another kind: 0.2 for 240 and 0.8 for zero.

[10] H.H.Gossen, Entwicklung der Gesetze des menschlichen Verkehrs und der daraus flieszenden Regeln für menschliches Handeln,1853, 3. Auflage, Prager, Berlin, 1927,6.

[11] V.Pareto, *Manuale di Economia Politica*,1906, translation revised by Pareto himself:: *Manuel d'économie politique*, second ed., Paris,1927.

[12] H.Gossen,, *The Laws of Human Relations and the Rules of Human Action Derived Therefrom*,1853, translated by R.C.Blitz, The MIT Press, Cambridge (Mass), 1983, 9.Edition in German of 1927,6.

[13] W.Vickrey, *Microstatics*, Harcourt and Brace, New York - Chicago, 1964.

[14] G.Stigler, *Notes on the history of the Giffen Paradox*, Journal of Political Economy, 152-155, April 1947.

[15] H.Minkowsky, *Geometrie der Zahlen*, Teubner, Leipzig,1896,35.
D.Hilbert, *Gesammelte Abhandlungen von Hermann Minkowski*,1911, Teubner, Leipzig, reprinted by Chelsea publishing company, New York, 1967,XI and Vol.II,103.
A.V.Pogorelov, *The Minkowski Multidimensional Problem,* John Wiley, New York, 1978, 9.

[16] For this point, see: W.W.Leontief, *The Use of Indifference Curves in the Analysis of Foreign Trade*, Quarterly Journal of Economics, 1933, 493-503

[17] W.W.Leontief, *op. cit.* More developed in: C.P.Kindleberger, *International Economics*, first edition, Irwin Inc. Homewood (Ill.), 1953,102-103.

[18] Translation from French by the author after the second French edition (1906) 1927, Pareto, *op.cit.*354.

CHAPTER 4

THE INDIFFERENCE METHOD

The Indifference Method has two stages: the filtering stage followed by the indifference stage sensu stricto[1].

4.1 The Filtering Stage of the Indifference Method

Some of the candidate alternatives are excluded if they do not respond to conditions concerning lower bounds or upper limits in a filtering stage (cf. Part III, 3.2).

4.2 The Indifference Stage of the Indifference Method

Let us return to the pater familias buying a car. Cars B and C have excellent comfort, fast speed and a special shape. At this point, the pater familias is indifferent between cars B and C. However, the price will finally decide: $40,000 for car B and $35,000 for car C.

In the indifference approach the vector of attributes of a candidate alternative z is called indifferent towards the vector of attributes of another candidate alternative y, if: $z_i \approx y_i$ for all i, with a single exception in which: $z_i > y_i$; z_i represents attribute i attainable in alternative z and y_i attribute i attainable in alternative y $(i = 1,2,...,n;$ n the number of attributes).

On this basis: $z\,P\,y\,P\,x\,P\,u$ is also possible iff $z_i > y_i > x_i > u_i$ for just one i with indifference for the other attributes.

Comparing the alternatives, the stakeholders are indifferent for all objectives except one. This only one exception mostly belongs to the cost sphere in the set of objectives. Indeed in real life some stakeholders are indifferent about alternatives when some thresholds or ceilings are respected. Most of the commodities and services however are no free commodities or services and in the traditional scarce economy the cost side takes then a decisive position. Therefore, in the Indifference Stage, when candidate alternatives have passed the filtering stage and the stakeholders are indifferent about the alternatives, the cost side will decide on the optimal choice to be made.

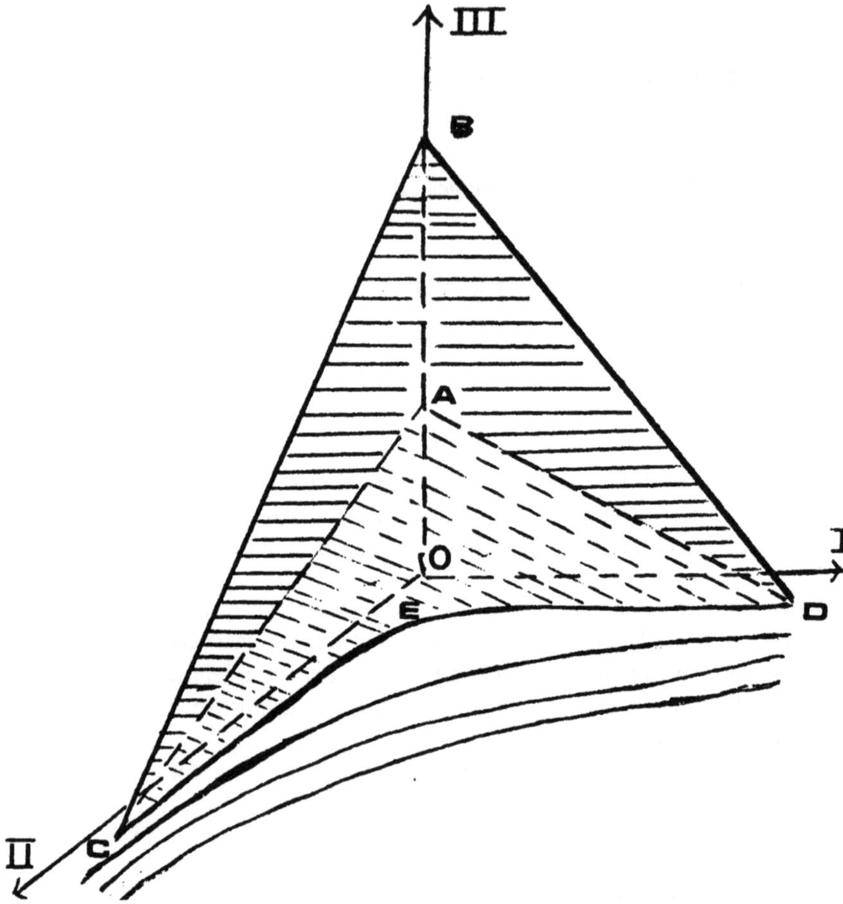

Fig. 4.1 Indifference Method with three Objectives and two Alternatives

The picture of the cost side will change according to the nature of the stakeholders, such as private entrepreneurs, civil servants or a government. Private entrepreneurs will see costs as a cost-revenue scheme, i.e., they will aim at the maximization of *Net Present Value* (NPV) or of *Internal Rate of Return* (IRR). Civil servants looking after the budget will strive for minimization of the discounted budget costs. Governments, considering themselves as the protectors of the general interest, will rather strive after the maximization of the Discounted Value Added or even better the *Discounted Social Value Added*. Indeed if well-being economics rather than welfare economics is aimed at, i.e., including also social costs (like pollution) and social benefits, the corresponding maximization would be called the maximization of *Social Value Added* (SVA).

Let us consider the simpler case of cost minimization with only two objectives and two alternatives. The candidate alternatives, which were able to pass the filtering stage and for which the stakeholders looking for utility are indifferent, form then a part of a single indifference curve which is determined by the feasibility, due to the available resources e.g., labor and capital, of the two objectives (Fig.4.1).

Every point on the curve CED is assumed to have the same utility towards the objectives I and II. In fact the budget cost will form a third objective, this time to be minimized. In case of two alternatives the alternative corresponding to the plane ACED, which is situated lower than the plane BCED of the second alternative, is preferred. It has to be noted that the third objective, the cost objective, contrary to the two other objectives, is considered in a linear way, which is a general assumption for the cost objective in indifference curve analysis. Indeed, for Vickrey[2] the money side is regarded as a very broadly defined commodity with usually no saturation point, or at least if such a point exists it will be far beyond the scope of any normal diagram.

The indifference method is only applicable for the *Discrete* and not for the *Continuous Case.*

General agreement on indifference between the stakeholders may be difficult, certainly when many stakeholders and many objectives are involved.

It is clear that the indifference method is only applicable due to a special attitude of the stakeholders. Therefore, it is better to look for a more general applicable approach[3].

CHAPTER 4 - ANNEX 1

CASE STUDY: THE INDIFFERENCE METHOD FOR ARMS PROCUREMENT

The Indifference Approach was used in Belgium for procurement of imported heavy military equipment such as rockets, tactical aircraft, escort vessels and heavy trucks[4].

Previously the difference between budgetary costs and macro-costs had to be explained. Budgetary costs cover the direct costs, which are charged on the budget of the department of defense. Macro-costs are equal to the budgetary costs minus social benefits such as increase in tax revenue and reduction of unemployment allowances. This reduction is the case when new jobs are created in a situation of unemployment. If an alternative is chosen with lower macro-costs but with higher budgetary costs than another alternative, the difference in budgetary costs has to be charged to another budget than the military one e.g., on the budget of the department of economic affairs or of finance.

More in detail, the procurement of heavy trucks of six tons will be explained, with especially details on the hard constraints. The candidate trucks had to undergo heavy tests: to climb slopes of 70%, to drive through 1.60m of water, to stay in water for some time, to overcome a lot of obstacles, to use consecutively petrol and light fuel, etc. The characteristics of the trucks were written down before and after the tests. Finally the military specialists were indifferent among a limited number of types.

It has to be noted that very often military specialists put their aspiration levels too high. Solutions are then mostly far away from their aspiration levels. They have to be satisfied by being indifferent about alternatives, which already passed a severe filtering stage.

In a next step economists will interfere and in that case the lowest macrocost will decide the choice.

The problem is a bit more difficult for equipment still in a development process. At that moment a feasible domain, which is based on applied research, is delimitated in order to generate alternatives during the process. Even then, the procedure, which was described above, is applied.

Nowadays the Belgian national defense attitude is different. We shall come back to this point later.

This indifference approach has the intention to solve the problem of optimizing several objectives if several alternatives are possible and the stakeholders are indifferent about the objectives but not about the costs. A severe filtering stage will precede an indifference attitude.

An Indifference Approach where the stakeholders are indifferent between the alternatives concerning the objectives with exception of the cost side is certainly also applicable for the private sector. Real world applications, however, are rather known for the government sector such as for national defense.

NOTES PART 3 CHAPTER 4

1 The Indifference Method was invented by the author: see M.Despontin, J.Moscarola, J.Spronk, *A user-oriented listing of multiple criteria decision methods,* Revue belge de Statistique, d'Informatique et de Recherche Opérationnelle, vol 23,4, December 1983, 20.

2 W.Vickrey, *Microstatics*, Harcourt and Brace, New York - Chicago, 1964.

3 This could also be a reaction against what Schärlig calls "Methods of Partial Aggregation", such as the several Electre methods, Prométhée, Qualiflex and Oreste (A.Schärlig, *Décider sur plusieurs critères*, Presses Po

lytechniques Romandes, Lausanne, 1985,139-231.

4 W.K.Brauers, *Multiple Criteria Decision Making and the Input-Output Model for National Defense*, Belgian Department of National Defense, Brussels,1977.

W.K.Brauers *The Indifference Approach in Multiple Criteria Decision Making,* Research Paper N° 5, University of Antwerp (RUCA), Institute for Developing Countries,1988.

W.K.Brauers *Some Principles of Multi-Optimization*, Systems Sciences and Engineering, Edited by Cheng Wei-Min, International Academic Publishers and Pergamon, Beijing(Ch.),1988, 678-682.

W.K.Brauers, *Our Army is awake!* (in Dutch: ons leger waakt ! waakt ons leger wel?) de Vries-Brouwers, Antwerp,1993.

CHAPTER 5

HOW TO OPTIMIZE IN A STAKEHOLDER SOCIETY?

Humans have many aspirations, which we translate as "Objectives". A problem arises if these objectives are independent or quasi independent from one another. Moreover, individual objectives have to agree with the aspirations of a Stakeholder Society.

The time has come to make some preliminary conclusions about the optimality of these objectives in a Stakeholder Society.

1) **In the past multiobjective optimization acted by a decision-maker-dictator, assisted by an adviser-researcher. Nowadays, especially knowledgeable people are the indirect representatives of the stakeholders by parliament, consumer organizations, trade unions or associations of employers, which on their turn are assisted by study centers. Finally, if there is a principal he will take a final decision.**

2) **For any problem, solving the number of Independent Objectives has to be limited as much as possible. Therefore, Interdependence, Dependence and Quasi-Interdependence Relations are helpful** (see Part I, 2.7 and 2.8)

   Quasi-Interdependency means a rather vague notion of interdependence as observed for instance in nominal methods, simulation and scenario writing.

   Some methods of Creative and Prospective Thinking, such as Delphi, Nominal Group Technique, Cross-Impact, Simulation and Scenario Writing, are very helpful for assistance in observation of all kind of Group Phenomena, such as in a Stakeholder Society (see Part II).

3) **In addition, Hierarchy of Objectives, either as a Complete A Priori Priority or as a Pyramid of Objectives, has to be taken into account** (see Part III, 1.6 and 1.7).

4) **The Utility of Objectives is expressed by Attributes, which are measured in a direct or indirect way. Even Quality has to respond to that rule. In borderline cases this utility could be represented by a dimensionless number, but not by a Nominal Scale such as: bad, moderate, good** (see Part III, Chapter 1).

   The Attributes are measured for several alternative solutions, which are simply called "Alternatives".

5) **Weighting of Objectives has to be avoided, which creates in fact a *Superobjective*** (see Part III, 2.4).

6) **Decreasing Marginal Utility and Indifference Curve Analysis suggest that** (see Part III, 3.4):

   - **methods based on linearity have to be omitted.**
     In a firm, however, the formula of convex linear combination can be used under the assumption of constant returns to scale with a rigid production function (cf. the fixed technical coefficients in the input-output production functions, see Part I, 2.4). Soon or later, productivity changes will appear at the supply side, whereas finally decreasing marginal utility at the demand side will interfere (see Part III, 2.4);
   - **for a given set of data, several Pareto Optima are possible simultaneously.**

7) **Two by two comparisons of alternatives may lead to reverse orders and even to contradictory results. Instead, multifarious comparisons are made** (see Part III, 2.5).

8) **If multifarious comparisons of alternatives are made some will drop out when hard constraints are set in a filtering process. The remaining ones can be ranked by dominance or by indifference. These procedures do not always work for the comparisons of alternatives. A set of until now incomparable alternatives will remain** (Part III, Chapters 3 and 4).

   For this purpose, Indifference Curves and certainly Community Indifference Curves are difficult to draw in practice. To find methods as substitutes is the task of the following parts of this book.

PART FOUR

REFERENCE POINT THEORY

A METHOD TO COMPARE

UNTIL NOW INCOMPARABLE

ALTERNATIVES

CHAPTER 1

REFERENCE POINT THEORY AS A SOLUTION FOR MULTIOBJECTIVE UTILITY

Distinction has to be made between the continuous and the discrete cases. All continuous cases will provide at least one solution.

Problems may arise with the discrete cases, with ranking as the outcome. Indeed, a ranking is still missing for the candidate alternatives, which are neither strongly nor weakly dominating, nor being entirely dominated. A ranking is also absent if there is neither indifference in the choice of the alternatives nor an a priori priority of objectives. A *Set of Until Now Incomparable Alternatives* finally remains.

1.1 THE RANKING STAGE FOR THE UNTIL NOW INCOMPARABLE ALTERNATIVES

Let us return to our pater familias, who was already cited in the previous parts of the book. He wanted to buy a new car. He had the choice between five cars: A, B, C, D and E. The objective "comfort" possesses the following attributes: excellent, medium and weak. The comfort of car E is weak. A weak comfort was in any way not acceptable for the pater familias. Consequently car E does not pass the filtering stage and the weak comfort forms a lower bound: the comfort has to be better than weak, namely medium or excellent. Car D costs \$75,000. The budget set by the pater familias was a maximum of \$50,000. The amount of \$50,000 is an upper limit and car D was no longer considered. He has still the choice between cars A, B and C.

Car A showed the following attributes:

price: \$45,000
comfort: excellent
speed: medium
shape: ordinary

Car B:

price: \$40,000
comfort: excellent
speed: fast
shape: special

Car C:

price: 35,000$
comfort: excellent
speed: fast
shape: special

Our pater familias chooses car C. Car C dominated car B and B dominated A. In this case, there is *Transitivity:* C also dominated A.

Suppose that an additional car F becomes available, with the following characteristics:

price: \$30,000
comfort: excellent
speed: medium
shape: special

Until now his preference went to car C:

price: \$35,000
comfort: excellent
speed: fast
shape: special

These two alternatives are incomparable and our friend does not know what to do. A solution for this problem has to be found.

1.2 GENERALIZATION FOR THE UNTIL NOW INCOMPARABLE ALTERNATIVES

The vectors of attributes of candidate alternatives z and y are until now incomparable to each other, if and only if: $z_i \geq y_i$ for some i and $z_i \leq y_i$ for some other i, with $i = 1, 2,.., n;$ n being the number of attributes.

Ipso facto candidate alternatives z and y are represented in *Euclidean Space* by points or respectively by:

$$z = (z_1, z_2,..,z_i,..,z_n) \text{ and}$$
$$y = (y_1, y_2,..,y_i,..,y_n).$$

A point is then called *Efficient* if some of its co-ordinates are nondominated by the corresponding co-ordinates of other points[1].

The elaboration of a ranking for the until now incomparable alternatives or for efficient points forms the core of the theory presented here. Linearity with weights would infringe the economic law of decreasing marginal utility. In addition, linearity creates a *Superobjective*, which ipso facto is mono-objective. Piecewise linearity approximates decreasing marginal utility, but for each piece, weights are used and their total creates again a superobjective.

Any theory on multiobjective utility theory has to take into account the law of decreasing marginal utility and consequently indifference curve analysis. However, indifference curves are difficult to determine.

A theory called *Reference Point Theory* seems to be an acceptable substitute. In Reference Point Theory the co-ordinates of a reference point are determined after strict rules. Using Reference Point Theory signifies too that the attributes have to be normalized. Indeed due to different units of measurement a *Nonadditive Normalization* for the attributes has to occur, which means to give a score to each attribute. The adjective "nonadditive" indicates the non-necessity of addition of the scores, as distinct from weights.

How to measure the distance between the co-ordinates of the alternatives and of the Reference Point, forms another problem to be solved.

1.3 INDIFFERENCE CURVE ANALYSIS AND REFERENCE POINT THEORY

Indifference Curves Analysis solves in fact the problem of the until now incomparable alternatives. However, as was already mentioned, indifference curves are extremely difficult to draw. Do there exist acceptable substitutes? Therefore we make the connection between indifference curves and reference theory.

Reference Point Theory[2] is compatible with the law of decreasing marginal utility, with nonlinearity, with Pareto optima and with the consideration of discrete points. Consequently this theory is selected to solve the problem of the ranking for the until now incomparable alternatives.

Suppose three alternatives, which are represented by three discrete points A, B and C, facing two objectives translated into two attributes. Imaginary indifference curves favoring C are drawn in full lines and imaginary indifference curves favoring A are drawn in dotted lines (Fig.1.1)

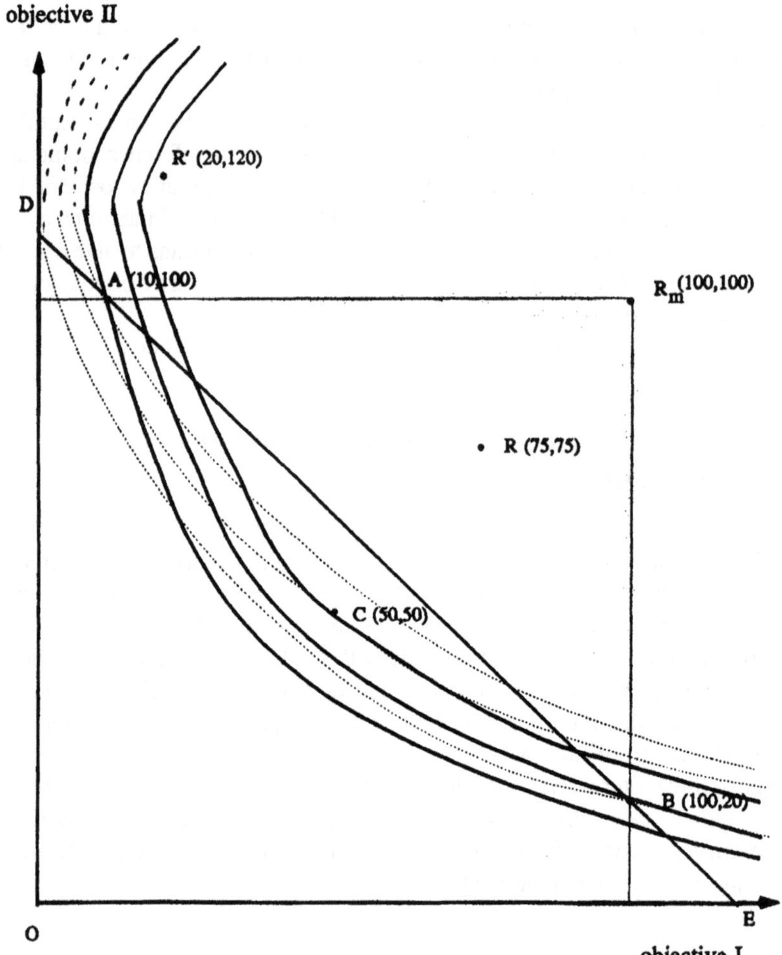

Fig. 1.1 Discrete Points, Imaginary Indifference Curves and Reference Points

With the *Additive Linear Method with Weights* an in-between alternative like C (50;50) will never be ranked first before A (10;100) and B (100;20), even when a higher preference is given to attribute I, or when a higher preference is given to attribute II, or for any weights given to attributes I and II.

Example on basis of Fig.1.1

$$w_I x_I + w_{II} x_{II}$$

with x_I : attribute I
x_{II}: attribute II

1. Same weights

$$w_I = w_{II} = 0.5$$

A = 55; B = 60; C = 50

B P A P C

2. Higher weight for objective I

$$w_I = 0.9;\ w_{II} = 0.1$$

A = 19; B = 92; C = 50

B P C P A

3. Higher weight for objective II

$$w_I = 0.1;\ w_{II} = 0.9$$

A = 91; B = 28; C = 50

A P C P B.

Decreasing marginal utility could, as contrasted with additive linear methods with weights, favor C (50;50) above A (10;100) and B (100;20):

C P A

C P B

Imaginary nonlinear indifference curves favoring C may give the following comparable ranking (see Fig 1.1):

C P B P A.

In its turn, Reference Point Theory has no problem in arriving at the same conclusions. Indeed if the stakeholders agree about a Reference Point, for instance R (75;75), which is in favor of C, the ranking based on straight lines between the discrete points and the reference point is once again:

C P B P A.

Imaginary Indifference Curves, drawn in such a way that they are favoring A, may give the following ranking (see fig 1.1):

A P C P B.

If the stakeholders are in favor of A, then they could choose for instance reference point R'(20;120) leading once again to the straight ranking :

A P C P B.

Consequently, the final results from reference points are comparable with those of decreasing marginal utility and of indifference curve analysis even for discrete points.

However, several problems remain:

1) the choice of the reference point if several discrete points are given;
2) the normalization of the attributes;
3) the measurement of the distances between the discrete points and the reference point;
4) the danger of reversing the order for the alternatives
5) and finally the definition of a reference point for alternatives, which are generated during the process. Indeed a distinction has to be made between the cases where alternatives are already present and the cases where they are generated.

1.4 CONSIDERATION OF THE MAXIMAL CRITERION REFERENCE POINT

The reference points which were mentioned until now were arbitrarily chosen in a subjective and even utopian way. Therefore, preference is given to a reference point possessing as co-ordinates the dominating co-ordinates per attribute of the candidate alternatives and which is designated as the *Maximal Criterion Reference Point*. This approach is called realistic and non-subjective as the co-ordinates, which are selected for the reference point, are realized in one of the candidate alternatives. In our example (see Figure 1.1 in the previous section), the alternatives A (10;100), B (100;20) and C (50;50) will result in the maximal criterion reference point R_m (100;100).

More than two attributes and objectives are represented by way of a vector belonging to the n-dimensional Euclidean vector space Rn. In the case of *m* alternatives the following matrix is obtained :

$[x_{ij}]$

with : $i = 1,2,\ldots\ldots,n$ as the attributes
$j = 1,2,\ldots\ldots\ldots,m$ as the alternatives.

Under these circumstances, a vector called the *Maximal Criterion Vector* also represents the reference point in question. This Maximal Criterion Vector $\{r_i\}$ is composed of the co-ordinates of the reference point and is represented by a column vector:

$\{r_1, r_2, \ldots, r_i, \ldots, r_n\}$

with *n* attributes.

The Maximal Criterion Vector is self-evident, if the alternatives are well defined, as for projects in Project Analysis and Project Planning. However, some call this realism only a *Satisficing Result* or *Bounded Rationality* as it seems that the stakeholders are completely satisfied if the realistic reference point is reached[3].

As mentioned, the Maximal Criterion Vector is composed of co-ordinates which are realistic. This is not the case with the *Utopian Criterion Vector*, or with the *Aspiration Criterion Vector*.

Contrary to the maximal criterion vector, an *Ideal Point* may have co-ordinates, which are freely chosen by the stakeholders, either under the form of a Utopian Criterion Vector, or of an Aspiration Criterion Vector.

1.5 THE UTOPIAN CRITERION VECTOR AND THE ASPIRATION CRITERION VECTOR

The co-ordinates of a *Utopian Criterion Vector* are formed as follows[4]: $r^*_i = r_i + \varepsilon_i$

The utopian criterion vector gives higher values to the co-ordinates of the reference point than the maximal criterion vector. This is understandable for instance for *Performance Management*, such as for student evaluation and for any performance in the private and the public sector.

A desirable reference point as in the utopian criterion vector method is defendable, though the reference point in question is determined in a subjective way.

The co-ordinates q_i of an *Aspiration Criterion Vector* are formed as:

$$q_i \leq r_i$$

$(r_i - q_i)$ being a subjective element.

The aspiration criterion vector moderates the aspirations by choosing smaller co-ordinates than in the maximal criterion vector. Indeed stakeholders may be more moderate in their expectations.

The utopian criterion vector, more than the maximal criterion vector or the aspiration criterion vector, offers a better response to the idea of *Multiobjective Optimization*. Indeed, in theory the size of a positive ε_i; is unlimited.

1.6 THE NORMALIZATION OF ATTRIBUTES IN REFERENCE POINT THEORY

In order to use reference point theory the attributes are normalized. Indeed due to different units of measurement, a *Nonadditive Normalization* for the attributes has to occur. Normalization means then that the stakeholders are asked, with the assistance of expert techniques such as Delphi, to give a score to each attribute. The adjective "Nonadditive" indicates the non-necessity of addition of the scores.

Minima are set over into maxima by allocating the highest scores to the alternatives with the lowest figures of the attributes.

Mechanical methods are also proposed for normalization[5]. In the mechanical methods, there is no interference of stakeholders or of experts. The methods can be used as a neutral outside norm.

Normalization procedures are quite common in econometrics and in mathematics for economists. In factor analysis the procedure is used in order to calculate the deviations from the arithmetic mean and the standard deviation. In input-output analysis, normalization of the technical structure into technical coefficients occurs in order to invert the technological matrix.

What is the difference between scores of the attributes in Reference Point Theory and weights, so much in use in other methods considering multiple objectives? We follow therefore the definition of Schlaifer[6] concerning weights, as was mentioned above. His rules stress the additive property of weights.

For instance institutions like Moody's, Standard and Poor's et al. make ratings which are based on giving weights to solvability, to provisions for bad debtors, to return on equity and to productivity in order to satisfy the needs of a passive investor.

Weighting procedures in Multiobjective Utility Theory leads to mono-objectivity under the form of a single super-utility concept. On the contrary, *Scores* for attributes in Reference Point Theory are parts of row vectors:

$$x_j = [s_1x_{1j},\ s_2x_{2j}, \ldots,\ s_ix_{ij}, \ldots\ s_nx_{nj}]$$

with: s_i = the score of attribute i ($i = 1,2,\ldots n$, n the number of attributes)
x_j = the row vector of utility for alternative j ($j = 1, 2,\ldots, m$; m the number of alternatives).

In this way a *Normalized Matrix* is obtained: $[{}_Nx_{ij}]$
N indicates the normalization
${}_Nx_{ij}$= the normalized attribute i of alternative j.

1.7 HOW TO MEASURE THE DISTANCE BETWEEN THE DISCRETE POINTS OF THE ALTERNATIVES AND THE REFERENCE POINT?

Previously in connection with Fig.1.1, straight lines were operationally drawn between every point of the alternatives and the Reference Point. Is this measure of distance acceptable? The objections against the linear distances will be formulated when the Euclidean distance metric is discussed below.

The *Minkowski Metric* as a discrepancy measure will bring however the most general synthesis[7]:

$$Min.\ M_j = \{\sum_{i=1}^{i=n} (r_i - {}_Nx_{ij})^{\alpha}\}^{1/\alpha}$$

with: M_j = Minkowski metric for alternative j
r_i = the ith co-ordinate of the reference point
${}_Nx_{ij}$ = the normalized attribute i of alternative j
$j = 1,2,\ldots,m$ with m as the number of alternatives
$i = 1,2,\ldots,n$ with n as the number of attributes

The Minkowski metric represents the basis of what is designated in the literature as *Goal Programming*[8]. From the Minkowski formula, the different forms of goal programming are deducted. The metric shows these forms depending on the values given to α.

With the *Rectangular Distance Metric* ($\alpha = 1$), the results are very unsatisfactory. Suppose e.g., the reference point (100;100) then the points (100;0), (0;100), (50;50), (60;40), (40;60), (30;70), and (70;30) all show the same rectangular distance and they all belong to the same line: $x + y = 100$. Ipso facto, a midway solution like (50;50) takes the same ranking as the extreme positions (100;0) and (0;100). In addition, the points: (30;30), (20;40), (40;20), (50;10), (25;35), (0;60) and (60;0), all belonging to the line: $x, + y = 60$, show the same rectangular distance to a reference point (50;40), which is not defendable.

In order to counterbalance this problem, weights could be given to the attributes[9]. In this way, once again, the negative remarks are involved that were made earlier concerning weights and linearity.

With $\alpha = 2$, radii of concentric circles, with the reference point as central point, will represent the Euclidean distance metric. This distance metric applied for two attributes is similar to the linear distances, which were cited above, However, as a convex outcome, it would not produce evidence for optimality of nonconvex indifference curves, to which the points may belong. Moreover, applying the Euclidean distance metric for the first example, which is given above, the outcome is very unusual. The midway solution (50;50) is ranked first with symmetry in ranking for the extreme positions: (100;0) and (0;100); the same for (60;40) and (40;60) and for (30;70) and (70;30).

The *Euclidean Distance Metric*, characterized by three attributes, is represented by radii of concentric spheres, with the reference point for center. This convex outcome does not produce evidence of optimality for nonconvex surfaces. By analogy, one could imagine that for more than three attributes corresponding convex manifolds would not bring an optimal outcome for manifolds, which could be nonconvex.

With $\alpha = 3$, negative results are possible if some co-ordinates of the alternatives exceed the corresponding co-ordinate of the reference point. It is also not clear if nonconvex solutions have a chance for optimality.

The same is the case with $\alpha > 3$, with the exception of $\alpha \rightarrow \infty$. In this special case of the Minkowski metric only one distance per point, viz. the largest one, is kept in the running. Using the Minkowsky metric in its most extreme ∞-norm, nonconvex outcomes can be obtained, as is the case with indifference curve analysis.

The Minkowski metric becomes a *Min-Max Metric*. If the following matrix is given:

$$[r_i - {}_N x_{ij}]$$

with: $i = 1, 2, \ldots, n$ as the attributes

$j = 1, 2,..., m$ as the alternatives
r_i = the ith co-ordinate of the reference point
${}_Nx_{ij}$ = the normalized attribute i of alternative j

then this matrix is subject to:

$$\underset{(j)}{Min}\ \{\underset{(i)}{max}\ (r_i - {}_Nx_{ij})\}$$

This *Distance Function* will form the basis for *Reference Point Theory* from now on.

Returning to the example, which is given in Fig.1.1 of section 1.3 of this chapter, the formula is applied as follows:

Table 1.1 Deviations from the Reference Point for two Attributes

Attributes	$r_i = R_m$	A {10;100}	B {100;20}	C {50;50}
I	100	90	0	50
II	100	0	80	50
Largest deviation		90	80	50

The following ranking is the result: C P B P A

If hard constraints of the filtering stage have to be included, then the limits of lower bounds (thresholds) or upper limits (ceilings) have to precede the minimax metric. Beforehand the possibility of dominance has to be examined too.

1.8 THE PROBLEM OF AN EFFICIENT POINT THAT IS NOT PROPER EFFICIENT AND THE REFERENCE POINT THEORY

An efficient point that is proper efficient allows for a very large gain in one co-ordinate against a very small loss in another co-ordinate compared to an efficient point, which is not proper efficient[10]. Suppose for instance for two objectives the following efficient points: A{10;100}; C{50;50}; B{100;20} and F{200;45}. Point F compared to point C gains very much in one co-ordinate against a very small loss in another one. Point F is then called proper efficient but point C not (for an Efficient Point, see: Part IV, 1.2).

There is no problem arising for proper or non-proper efficiency in Reference Point Theory with a Maximal Criterion Vector. Indeed, for the example with the four efficient points the following deviations from the reference point are written down in Table 1.2.

Table 1.2 Deviations from the Reference Point for efficient points

r_i	A	B	C	F
200	190	100	150	0
100	0	80	50	55
Largest deviation	190	100	150	55

Minimizing the largest deviations gives the following ranking: F P B P C P A

Point C being a non-proper efficient point compared to point F ranks indeed low in the ranking of the four points.

1.9 IS THERE A THE DANGER OF A REVERSE ORDER?

Can a reverse order in ranking, which is suggested by Arrow's General Impossibility Theorem[11], occur with a Reference Point?

The danger is real for methods distinguishing series of two by two comparisons, which are brought together in a multilateral setting afterwards. It is reasonable that stakeholders themselves may change their priorities by switching from the Reference Point, which is obtained through the Maximal Criterion Vector Method, to a more desirable point with a reverse order in ranking as a result. If a sufficiently large number of alternatives is present the stakeholders may wish to change the first outcome by changing one of the co-ordinates of the reference point and if not satisfied with the second outcome they may switch to another co-ordinate etc. till a satisfactory outcome is obtained (*Interactive Method*). It is also acceptable that the stakeholders change their opinion after some time with a reverse order in ranking as a result.

A reverse order in ranking which is caused outside the stakeholders' opinion however is not acceptable. It could happen with the Maximal Criterion Vector Method if an additional alternative is introduced afterwards.

At that moment the following situations may occur. First, the co-ordinates of the new alternative are smaller than those of the reference point and consequently no problem arises. There is only a problem if one or more co-ordinates of the new alternative are larger than those of the existing reference point. A reversed order in the preferences for the alternatives is then possible. In order to avoid this possibility it is advisable to maintain the old reference point, but then negative distances may arise.

Suppose e.g., the following reference point: r* (15,000;6,500;400) and a new alternative: M (15,000;12,000;0), then the deviation for the second co-ordinate would be: - 5,500. Therefore, absolute values have to be introduced in the min-max metric:

$$\underset{(j)}{Min}\{\underset{(i)}{max}\,|r_i - {}_Nx_{ij}|\}$$

with: $i = 1, 2,..., n$ as the attributes
$j = 1, 2,..., m$ as the alternatives
r_i= the ith co-ordinate of the reference point
${}_Nx_{ij}$= the normalized attribute i of alternative j.

1.10 ALTERNATIVES GENERATED DURING THE PROCESS: THE CONTINUOUS CASE

Until now, the alternatives were already known in detail beforehand. It is also possible to generate alternatives during the process itself. Hwang and Masud[12] distinguish therefore two very different categories: Multiple Attribute Decision Making (MADM) and Multiple Objective Decision Making (MODM). MADM in their view is associated with selection problems, MODM with design problems. In their book they maintain that previously major emphasis has been put on MADM, whereas they focus their attention rather on MODM. We do not think such a clear-cut distinction exists between the two categories. At first sight, it seems impossible to determine a maximal criterion reference point for MODM, but two possible scenarios may present an outcome.

First, the stakeholders fix hard constraints and for the remainder they are indifferent with the exception of the cost problem. Logically there will be a lower bound on the micro or macro cost. This lower bound will determine the solution. In the absence of this indifference, the stakeholders themselves will be asked to fix a reference point, which is called under these circumstances a *Desirable Reference Point*. Constraints will here also limit a domain of feasibility. Several possibilities may then occur.

1° The desirable reference point is located inside the feasible domain. At that moment the desirable reference point indicates the final solution unless the stakeholders prefer[2] to switch to a dominating point or to a point on the borderline.

2° The desirable reference point is located on the border of the feasible domain in case the constraints show a weak order (≤or≥). At that moment again the desirable reference point will indicate the final solution.

3° The desirable reference point is located outside the feasible domain. At that moment, an *Iterative Method* is applied.

As an example, two attributes, which are already normalized, are assumed:

x_1 as economic production to be maximized;
x_2 as pollution damage to be minimized.

A capacity limit in production exists: $x_1 \leq 26$
A necessary minimum of employment: $x_1 \geq 6$
A maximum on pollution allowed: $x_2 \leq 6$ (a pollutant already normalized).
The combination of production and pollution has to satisfy:

$$2x_1 - 6x_2 \geq 12$$

The stakeholders agree on a desirable point {19,7}, which is located outside the feasible region.

The min-max metric is applied for some points on the borderline, which are selected after an iterative procedure, initially based on numerous possible alternatives.

Table 1.3. Deviations from the Desirable Reference Point for some selected points

Desirable Reference Point	A{24;6}	B{21;5}	C{22;5 1/3}	D{23;5 2/3}	E{18;4}
19	5	2	3	4	1
7	1	2	1 2/3	1 1/3	3
Largest deviation	5	2	3	4	3

Preference: B P C I E P D P A

Once again a midway alternative, like B, wins.

1.11 CONCLUSIONS

This study had the intention to find solutions for the problem of optimizing several objectives, if several alternatives are present. Therefore, different approaches were discussed.

The weighting concept, in its linear or multilinear presentation, is rejected as in fact a single *Superobjective* is created. The concept which is based on a series of binary comparisons may often lead to reverse orders or to contradictory results. A strong and complete priority order of the objectives, which is dictated by the stakeholders beforehand, in fact boils down to a mono-optimization by stages. Each linear approach is rejected on the ground of the law of decreasing marginal utility. *Piecewise Linearity* approximates decreasing marginal utility, but is rejected too as each piece uses weights and the total of the pieces creates again a superobjective.

An indifference approach is useful but only applicable in limited cases, namely when stakeholders are indifferent between the possible alternatives concerning the objectives with exception of the cost side.

Finally, a general *Multiple Objective Utility Theory*, leading to *Multiobjective Optimization*, is proposed, which has the following stages:

1) a filtering stage in order to obey a set of hard constraints;
2) a ranking which is based on strong or weak dominance;
3) a ranking based on indifference curve analysis.

However, indifference curves, surfaces and manifolds are difficult to design in practice. Therefore, a substitute is proposed respecting decreasing marginal utility, nonlinearity, Pareto optima, reverse order problems and even nonconvexity.

This substitute will first exclude candidate alternatives, which cannot face certain hard constraints. Indeed the vector of attributes of each candidate alternative is compared to the vectors of the hard constraints. Hard constraints may have the form of lower bounds (thresholds) or of upper limits (ceilings). Next, the consideration of strong or weak dominance may already solve the problem. Otherwise, anyway for a discrete case, the remaining candidate alternatives are ranked with the assistance of the Minkowski Minimax Metric in comparison with a Reference Point. In a continuous case the reference point is a desirable reference point. In a *Continuous Case* the reference point is a *Desirable Reference Point.*

Many applications of reference point theory are possible. Some will be discussed in the following case study and in the following chapters.

1.12 CASE STUDY: HOW TO CHOOSE A WIFE?

The following objectives were chosen:

objective 1: max. cooking and household
objective 2: max. beauty
objective 3: max. intelligence

with the same attributes for cooking-household, beauty and intelligence:

1 (extra bad)
2 (bad)
4 (moderate)
8 (high)
16 (extra high)

The jury attributed the following points for candidates A to D:

Table 1.4 Attributes for Competing Candidates

Candidates	A	B	C	D
Attribute 1	1	16	4	2
Attribute 2	16	2	4	2
Attribute 3	2	1	1	8
Totals	19	19	9	12

No candidate dominates strongly or weakly the other candidates.
Candidate A is extremely beautiful, but poorly intelligent and extremely bad in cooking.
Candidate B is extremely good in cooking, but not at all beautiful and extremely bad in intelligence.
Candidate C is an average cook and an average beauty, but extremely bad in intelligence.
Candidate D is quite intelligent, but not at all beautiful and not good in cooking.

If the same weights are given to the three attributes and looking at the totals, the weighting method ranks as follows:

A I B P D P C.

In the weighting method, notwithstanding the distribution of the weights, the extreme positions will always win and the in-between alternatives will never have a chance.

The outcome of the Reference Point Method is different.

Table 1.5 Deviations from the Reference Point with the Maximal Criterion Vector

r_i	A	B	C	D
16	15	0	12	14
16	0	14	12	14
8	6	7	7	0
Max.	15	14	12	14

$$\underset{(j)}{Min}\{\underset{(i)}{max}(r_i - {}_Nx_{ij})\}$$

The final ranking: C P B I D P A means that an in-between candidate like C has a chance.

Only as a test the Euclidean Distance Metric is calculated on the maximal criterion vector:

$$\sqrt{\sum_{i=1}^{i=n}(r_i - x_i)^2}$$

$$A = \sqrt{15^2 + 6^2} = 16.16$$

$$B = \sqrt{14^2 + 7^2} = 15.65$$

$$C = \sqrt{12^2 + 12^2 + 7^2} = 18.36$$

$$D = \sqrt{14^2 + 14^2} = 19.80$$

ranking: B P A P C P D

This ranking is in complete disaccord with the previous results. The outcome confirms the theoretical demonstration, which was given previously in this chapter, namely that the Euclidean Distance Metric moves in other directions as the ∞ - norm of the Minkowsky metric. In addition, the Euclidean Distance Metric does not bring an optimal outcome for manifolds, which could be nonconvex. Therefore, we shall drop any calculation with the Euclidean Distance Metric from now on.

The Maximal Criterion Vector is a realistic approach. This is very important, when the alternatives are well defined as for projects in Project Analysis and Project Planning.

For *Performance Management*, as in the case of the marriage candidates, for students′ evaluation and for any performance in the private and the public sector, a desirable reference point as in the *Utopian Criterion Vector* method is defendable. However, the reference point in question is determined in a subjective way, unless the highest possible level is aimed at. In the case of the marriage candidates, for instance, the highest possible level, namely the reference point with co-ordinates: (16,16,16), is chosen. The score of 16 for intelligence determines a reference point, which is really utopian.

Table 1.6 Deviations from a Reference Point with a Utopian Criterion Vector

r_i	A	B	C	D
16	15	0	12	14
16	0	14	12	14
16	14	15	15	8
Max.	15	15	15	14

with as result: D P C I B I A

The most intellectual candidate D takes advantage of the increase in the intellectual co-ordinate of the reference point. In fact, D benefits from her extreme position.

NOTES PART 4 CHAPTER 1

[1] Efficiency in MOUT should not be confused with efficiency in cost-effectiveness analysis, where it means "with a minimum of costs" (see therefore: Part III, 2.2).

[2] R. Benayoun, J. De Montgolfier, J. Tergny and O. Larichev, *Linear Programming with Multiple Objective Functions: Step Method (STEM)*, Mathematical Programming, 1971 1, N°3, 366-375.
J.S.Dyer, *Interactive Goal Programming*, Management Science, 1972, 19 N°1, 62-70.
A.P. Wierzbicki, *Basic Properties of Scalarizing Functionals for Multiobjective Optimization*, Mathematische Operationsforschung und Statistik - Series Optimization, 1977, vol. 8, N°1, 55-60.
A.P.Wierzbicki, *The Use of Reference Objectives in Multiobjective Optimization*, Lecture Notes in Economics and Mathematical Systems, vol. 177, 1980.Springer-Verlag, Berlin, 468-486.
A.P.Wierzbicki, *A Mathematical Basis for Satisficing Decision Making*, Mathematical Modelling, vol. 3 N° 5, 1982, 391-405.
R.E.Steuer, E.Choo, *An interactive weighed Tcebycheff procedure for multiple objective programming*, Mathematical Programming,26,1983,326-344.
H.Nakayama, Y.Sawaragi, *Satisficing Trade-off Method for Multiobjective Programming*, Lecture Notes in Economics and Mathematical Systems, vol.229, 1984, 114-122.
R.E.Steuer, *Interactive Multiple Objective Programming: Concepts, Current Status and Future Directions*, International Conference on M.C.D.M. : Applications in Industry and Service, December 1989, A.I.T. Bangkok.
R.E. Steuer, Multiple Criteria Optimization: Theory, Computation and Application, Krieger Publishing Cy, Malabar (Fl.-U.S.),1989.

[3] A.P.Wierzbicki, *A Mathematical Basis for Satisficing Decision Making*, op.cit.
N.Ahituv, Y. Spector, *Evaluating Information under two Criteria: Expected Payoff and Risk*, Working Paper Nr 70/90, 1990, I.I.B.R., Tel Aviv University, Israel,5.

[4] R.E.Steuer, E.Choo, *op.cit.*

[5] A.Van Delft,P.Nijkamp, *Multi-criteria Analysis and Regional Decision-making*, M. Nijhoff, Leiden, Nl, 1977,37-38.

[6] R.Schlaifer., *Probability and Statistics for Business Decisions*, Mc Graw Hill, New York, 1959,8-11.

[7] H.Minkowsky, *Geometrie der Zahlen*, Teubner, Leipzig,1896.
H.Minkowsky, *Gesammelte Abhandlungen*, Teubner, Leipzig,1911.
D.Hilbert, *Gesammelte Abhandlungen von Hermann Minkowski*,1911, Teubner, Leipzig, reprinted by Chelsea publishing company, New York, 1967.

[8] Some, however, will also call it goal programming if the Lexicographic Method is determined as a function of the deviations of a goal. E.g., D.Giokas, M.Vassiloglou, A goal programming model for bank assets and liabilities management, European Journal of Operational Research, vol.50, N°1, 1991, 48-60.
Steuer develops a similar approach and calls it "preemptive goal programming". R.E.Steuer, Multiple Criteria Optimization: Theory, Computation and Application, Krieger, Malabar (Fla., U.S.), 1989, 292.

[9] M.Tamiz, D.F.Jones, *Goal Programming and Pareto efficiency*, Journal of Information and Optimization Sciences, vol.17 N°2, 1996, 1-17.

[10.] The notion of *Proper Efficiency* comes from M.Geoffrion, *Proper Efficient and the Theory of Vector Maximization*, Journal of Mathematical Analysis and Applications, Vol.22, 1968,.618-630.

[11] K.J.Arrow, *Social Choice and Individual Values*, Yale University Press, New Haven, 2nd. edit.,1963 (first ed. Wiley, New York,1951).

[12] C-L.Hwang, A.S.Masud, *Multiple Objective Decision Making; Methods and Applications,* Springer,, Lecture Notes in Mathematical Systems, Berlin, 1979, 6-7.

CHAPTER 2

PROJECT PLANNING, ANALYSIS AND EVALUATION

Project Planning or Project Analysis in entrepreneurial economics was originally a mono-objective decision making process, for instance with the consideration of Net Present Value or Internal Rate of Return[1]. It is true, however, that this mono-objective process is linked to a lot of uncertainty factors such as future inflation, interest rates and exchange rates. Moreover, in a stakeholder society it is no longer a mono-objective decision making process. Indeed at that moment, a lot of objectives have to be fulfilled such as: employment levels, balance of payments equilibrium, import substitution and export promotion, tax revenues and of course also net present value and internal rate of return. When an industrial project is brought at the level of government, either for nationalized enterprises or for private firms subject to a state license for operation, to public loans or to subsidies, these objectives will even be more enforced.

2.1 AN EXAMPLE OF PROJECT PLANNING

Some time ago, we submitted a proposal that was accepted by a government of a developing country. It concerned a proposal for an industrial development by a middle size firm. As the proposal was confidential, we can disclose neither the type of project, nor the country concerned. Therefore we shall speak of a production A in firm X in region Z of country Y, while the money terms are expressed in an imaginary currency, namely the Banco (B). Being interested in multiple objectives, we are only concerned about the technique and not about the specific environment in which the proposal was accepted. Indeed the proposal could be generalized for other projects and it will be shown how multiple objectives were involved. Of course, in the first place a technical knowledge about a specific production is necessary in order to propose this production. The technical specifications are not given here as not being relevant for the demonstration on multiple objectives decision-making. Moreover, it would disclose the kind of production concerned, which is not allowed.

2.2. THE OBJECTIVES FOR PROJECT PLANNING

At the money side, a high liquidity, a positive output measured by Net Present Value and Internal Rate of Return and an important income for government, have to be present. The liquidity objective means that at every moment the firm is able to pay its liabilities when due. The International Finance Corporation (IFC), affiliate of the World Bank for financing private investments in developing countries, is willing to finance a part of the assets on a long term basis at an interest rate of for instance, 13.71 %. It has to be proven that the criterion of an Internal Rate of Return higher than 13.71 % can be reached. At that moment, it is assumed that the government will participate with 50 % in the capital stock. However, the government will levy a corporate tax of 38 % and demands a total return on its capital stock of 100 % at the end of the period considered.

At the effectiveness side the fulfillment of an effective demand and positive effects in the regional employment level and on the national balance of payments are necessary Also several specific side conditions have to be fulfilled, viz., as it concerns: 1) a private firm; 2) a project for industrial development i.e., not agricultural or purely commercial. The government concerned is also interested in a production of commodities, which would create import substitution, i.e., the demand is already present but the commodities are imported. If the new product were exportable, it would mean an additional positive point. Export promotion may even come first. To this objective, the equilibrium of the balance of payments is linked, which may not be violated at any moment. Eventually a positive influence on the balance of payment is appreciated. The government is also interested in regional development and it has to be shown that e.g., 75 new jobs are directly or indirectly created in the region.

Nowadays, due to the uncertainty of the world economy, the lifetime of projects is shortened, to e.g., ten years, which does not mean that the production is not going on for a new period.

The government considered one single project without comparisons of projects or without putting them in competition with one another. In other words, until now, one project faces multiple objectives. Later on, the research is extended to different alternative projects.

2.3 THE SELECTION OF THE PROJECT

After the necessary feasibility studies on market and plant capacity, material inputs, location and site project engineering, manpower and project implementation, the following evaluation of investment and financing was made (Table 2.1).

Table 2.1 Evaluations of Investments and Financing in Banco (Year 0)

Assets	
Land	9,600 B
Buildings	120,000 B
Infrastructure	110,000 B
Factory equipment	842,600 B
Transport equipment	25,000 B
Costs occurred on all equipment	160,200 B
Costs of first installation	81,660 B
Office equipment	4,000 B
Other items	45,600 B
Permanent Working Capital	81,340 B
	1,480,000 B
Liabilities and Capital	
Capital Stock	
Promoter (no foreign nationality)	222,000 B
Part of government (foreseen)	222,000 B
Long Term Liabilities	
(World Bank at an interest rate of 13.71%)	828,000 B
Medium Term Liabilities	
(Participating bank)	208,000 B
	1,480,000 B

The Liquidity Objective

In the proposal, liquidity problems arise during the first three years. Indeed negative final cash flows occur in years 1, 2 and 3)[2]. Therefore, a contract was signed with bank V for a credit line of 300,000B.

The Objective of a Favorable Internal Rate of Return

The financial planning makes it clear that there is a supplementary need for working capital besides the working capital foreseen in the Investments and Financing Statement[3]. Consequently, this additional working capital has also to be taken into consideration in the computation of the Internal Rate of Return.

The lifetime of the project is foreseen as ten years, with a residual value of 600,700 B and with the possibility of extension of production after ten years. The break-even point is situated during the 10th year (net present value with 13.71 % interest rate at the end of the 9th year: minus 75,789 B; at the end of the 10th year: 30,313 B[4]).

The Internal Rate of Return of the Project is 16.08 %, much higher than the interest asked for by IFC. An interest rate of 16.08 % is certainly also satisfactory for the investors. The medium term loans are paid back before the sixth year and the World Bank loan is repaid by installments during the seventh, eight and ninth years (see Table 1.2 and [5]).

The Objective of Import Substitution and of Export Promotion

After thorough marketing research, it was found that[6]:

1) import substitution will be complete after three years;
2) the export rate will be maintained at 20 % from the third year on;
3) an important trade creation will take place in the country itself due to an adequate price policy and a distribution system all over the country.

Each year, a surplus of foreign currency becomes available totaling 1,722,000 B after ten years (see endnote 6). Even with an average depreciation of the local currency over ten years of 90 % acting on the imports, the foreign currency balance is still safe. Additionally, such depreciation would have a favorable influence on the level of exports.

The Objective of Regional Employment

Besides temporary foreign consultancy, regional employment is created for 97 persons divided as follows (Table 2.2):

Table 2.2 New jobs in Region Z of country Y[7]

Direct employment	72
Indirect employment	25
	97

The indirect employment was obtained by using the inverse matrix of the input-output tables of country Y.

The Objective of Government Revenue

The government asked for a total return on its investment of 100 % at the end of the period of 10 years. Reality was much better, namely 154%[8].

In fact, one has also to consider the income from taxation. At that moment, the total adds up to 1,184,780 B or 5.34 times the original investment (Table 2.3).

*Table 2.3 Total Government Income from Firm X over 10 years**

Government return on investment	341,563 B
38 % corporate tax	474,044 B
25 % tax on private dividends	65,803 B
40 % income tax on higher salaries	120,000 B
15 % indirect taxation on consumption from personnel	183,370 B
	1,184,780 B

* For details, see endnote 8.

Above the interest on the capital stock invested and taking into account a risk factor, government still obtains a net gain of 322,987 B i.e., a net return on its capital stock of 145 %[9].

As a conclusion, one may say that the project fulfilled all the objectives in a satisfactory way.

2.4 DIFFERENT COMPETING PROJECTS

Until now one project faced multiple objectives. Here the research is extended to different alternative projects.

With the use of the Indifference Method

Suppose several hard constraints, whereas one single objective, mostly at the cost side, has to be made optimum. We called this procedure: "the Indifference Method" (see Part III, Chapter 4).

Let us give a theoretical example. A government is willing to support a project in a region Z with maximization of the Net Present Value after 15 years, but the following hard constraints have to be fulfilled:

1) The break-even point at the side of the Net Present Value has to be reached before the 10th year;
2) the following guaranties on the balance of payments:

- during the first year no deficit on the balance of current account expressed in hard currency;
- during the second year: a surplus in hard currency on the current account of at least $500,000, to be repeated for the following eight years;

3) an investment in imported hard currency during the first two years of at least $1,250,000.

Suppose the following projects:

Project A

Project A guarantees a Net Present Value of $1,000,000 with a break-even point during the ninth year, fulfillment of condition two and an investment of $1,500,000.

Project B

Project B guarantees the same as project A but with a NPV of $800,000 and an investment of $2,000,000.

Project C

Project C guarantees a NPV of $825,000 with a break-even point during the ninth year. 2) A deficit in the balance of payments of $1,000,000 in hard currency is foreseen for the first year. In total during the following nine years a surplus in hard currency is foreseen of $5,000,000. 3) An investment of 1,250,000$ is planned.

Solution

A P B   Project C is excluded.

Different Objectives are foreseen besides different Hard Constraints

The following objectives are foreseen: maximization of Net Present Value, of employment, of the balance of payments surplus and of foreign investment. The projects give the same guarantees as above, plus for direct employment:

Project A: 500 jobs
Project B 1,000 jobs
Project C 1,000 jobs.

Only one hard constraint remains: the break-even point on the side of the Net Present Value has to be reached before the 10th year.

A Delphi exercise allots the following scores:

ten points for each 1,000 dollar NPV

one point for each 1,000 dollar for B. of, P. and investment

five points for each direct created job.

Table 2.4 brings a summary of all the attributes of the objectives.

Table 2.4 Attributes of Competing Projects

Attributes	Project A	Project B	Project C
Net Present Value	10,000	8,000	8,250
Direct Employment	2,500	5,000	5,000
1st year current account B. of P.	0	0	-1,000
Following years current account	4,500	4,500	5,000
Foreign Investment	1,500	2,000	1,250
Totals	18,000	19,500	18,500

If the scores are translated into additive weights, the following ranking is obtained after the totals of the table:

B P C P A

If a Reference Point Theory is used the result is different (Table 2.5).

Table 2.5 Deviations from the Reference Point with the Maximal Criterion Vector for Competing Projects A, B and C

Attributes	r_i	A	B	C
Net Present Value	10,000	0	2,000	1,750
Direct Employment	5,000	2,500	0	0
1st year current account B. of P.	0	0	0	1,000
Following years current account	5,000	500	500	0
Foreign Investment	2,000	500	0	750
Maximum deviation		2,500	2,000	1,750

After the Maximal Criterion Vector the following ranking is obtained:

C P B P A

This ranking is in disaccord with the weighting method, which anyway is rejected on the grounds which were mentioned above.

2.5 MULTIPLE OBJECTIVES IN INDUSTRIAL MANAGEMENT

A problem of renovation or replacement of old industry existed in the revival of Western Europe in the 1950s. Capital deepening or capital widening was a similar interpretation at that time.

The same dilemma was presented around 1994 in China. The Tiananmen Square uprising of May 1989 was not only a student's protest, but also a real revolution in which workers participated[10]. The following years saw the end of central planning in China. From that moment, universities and reformers of all kind started to think about alternatives such as renovation or replacement of old industry. In 1994, we had the opportunity to make some proposals in China, which we repeat here[11].

Several scenarios are possible, either based on updating of old industry or its replacement in the same sector or in other sectors. Their long-term period depends on the objectives e.g., ten or fifteen years. For the construction of a new river port or a seaport, at least 30 years have to be foreseen.

One objective is possible such as maximization of NPV, with constraints on pollution, imports of raw materials, energy etc. In the case of several objectives, such as maximization of NPV and of employment, minimization of imports, of raw materials, of pollution and of energy, Reference Point Theory is applied.. With that purpose in mind, a model is developed taking into consideration upper limits, lower bounds, dominating and nondominating effects, ending up with a set of nondominated alternatives, which are ranked by using the Reference Point Theory. In this way objectivity and the law of decreasing marginal utility are respected.

The case under consideration forms an illustration of the general tendency to substitute mono-objectivity by multiobjectives in micro- as well as in macro-economics.

1. Scenarios for solving the New-Old dilemma

Problems of Project Analysis

A project has to be considered in two phases: the feasibility and the execution phase. In the *Feasibility Phase,* several projects can be put into competition with one another. In the *Execution Phase,* either one or several projects can be realized. In the case of several projects, a final choice is made after a ranking, as a function of the available resources.

Inside the feasibility phase a distinction has to be made between the characteristics of the projects. In the case of *Independent Projects,* no interdependence exists between the projects. With *Contradictory Projects,* no common action seems possible, as stakeholders are entirely opposed to one another[12]. *Mutually Exclusive Projects* possess a broader definition. Besides contradictory projects they also include projects with an alternative time of beginning, limited projects funding or the case where one project is no longer required if the other one is carried out (e.g., the choice between a fuel thermal and an atomic power plant). Here the project is chosen with the highest rank classification. *Projects linked with essential Backward or Forward Related Projects* have to be taken as a totality. *Composite Projects* may operate either separately or as a group. In that case, the separate projects will be considered, whereas the sum of all the projects will be considered as a separate project. At that moment, synergy effects are possible: the sum of projects gives better results than the sum of the projects executed separately. Assuming three projects A, B and C, it would mean that (A+B+C) > A + B + C. In that case four projects will be taken into account: A, B, C and (A+B +C).

Scenarios with one Objective: the Maximization of Net Present Value

Two scenarios are considered: on the one hand the updating of old industry, replacement in the same sector or in other sectors on the other. Traditionally only a single objective is aimed at, namely the maximization of Net Present Value.

Is Net Present Value or Internal Rate of Return to be preferred? Certain disadvantages exist in the use of Internal Rate of Return as an objective (Table 2.6).

Table 2.6 Reactions of Projects against Net Present Value and Internal Rate of Return

Projects	Duration of the project	NPV	IRR
I	10 years	1000	18%
II	10 years	100	20%

If we take into consideration IRR, project II is chosen above I. If NPV is preferred, project I is chosen. Does there exist a contradiction?

If the potential lender or investor demands a rate of at least 15% (including a risk factor), then 15% becomes a lower limit of acceptability. In that case, project I is chosen. If the threshold is 19%, then project II is chosen. It means that IRR is acting as a constraint on the one hand and as a variable to be maximized on the other. In the Indifference Method or in the filtering stage, Net Present Value represents the variable to be maximized, whereas IRR acts as a constraint.

Finally, government, being the protectors of the general interest, will rather take the macro-economic point of view by for instance striving for maximization of the discounted Social Value Added (SVA)[13]. In this way, already several objectives are considered.

Scenarios with several objectives

In project analysis, several objectives can be considered. The private manager would think in the short run of maximization of his yearly turnover, of penetration in the national and international markets, of balancing his liquidity position, of minimizing his costs, the payment of taxes (minus subsidies) and the break-even period. In the long run, he aims to maximize the Net Present Value and the Internal Rate of Return and to minimize the Payback Period. The local government would like to maximize local employment, the revenue from local taxes and the redistributions in its favor from the national government, while the national government strives to maximize its tax revenue. Finally, macroeconomics, through the authority of the government or of supranational institutions, aim at maximization of direct and indirect employment, of the increase in Gross Domestic Product, of the surplus preferably in hard currency in the Balance of Payments. Therefore, we have to find methods which aim to satisfy several objectives simultaneously.

The Three Stages of the Reference Point Method

This more general method will have three stages. First in the filtering stage, alternatives to the objectives have to pass several constraints in the presence of lower bounds and upper limits. The second stage is based on dominance resulting in a ranking. The last stage is the stage for nondominated alternatives.

Due to the different units of measurement for the attributes of the objectives, *Normalization* has to occur. Therefore, the stakeholders are asked in a Delphi kind of operation to indicate a score for each objective.

The Execution Phase of Project Analysis

Once the feasibility phase has ended in a positive way[14], the stakeholders have to think about the execution process. After the available resources, an inductive method of realization is followed. With the available capital the stakeholders intend to realize one or more projects according to the ranking obtained from the feasibility phase. Liquidity considerations could already be included in the attributes of the feasibility phase.

On the macro-economic level, the total financing capacity could be limited. It could in its turn influence the level of the interest rate[15].

2. The Antagonism of the Old versus the New Industry in China

Let us give an example for the tooling industry. The permanent tooling is operated in a new factory, which is capital intensive resulting in a high NPV. The temporary tooling is done in an existing factory which is flexible enough to do this job but with smaller financial results. This example, which originally was mono-objective, could be extended to a multiobjective problem as stated in Table 2.7

After the available information, none of the projects seems to dominate strongly or weakly for all objectives (see Table 2.7, columns 3 and 4)[16]. Consequently, they form a set of nondominated alternatives.

Table 2.7 Reactions of Industrial Projects on the Proposed Objectives

(1) Attributes	(2) Thresholds or ceilings (filtering)	(3) Objectives	(4) Permanent Tooling	(5) Temporary Tooling	(6) Scores
Entrepreneurial economics					
1) NPV	≥ 0	max. in $	1.918 m. $	0.360 m. $	1 per 100 $
2) IRR	≥ 13%	max. in %	15%	14%	1,000 per interest point more
3) Payback period	5 years	min. in years	3 years	1 year	1,000 per year earlier
Government					
4) Total discounted government income	≥ 0.2 m. $	max. in $	0.5 m. $	0.2 m. $	1,000 per 100,000$ more
Macro-economics					
5)Direct employment	≥ 300 jobs	max. in jobs	300	2,000	10 per job
6) Discounted VA	≥ 100 m. $	max. in $	100 m. $	200 m. $	1 per 1,000$
7) Surplus in hard currency on current account Bal. of P. (discounted)	≥ 0.250 m. $	max. in $	1 m. $	0.250 m. $	1 per 1,000$

Column (3) of table 2.7 shows also the different units of measurement of the objectives. In order to *normalize* the objectives the stakeholders are asked in a Delphi exercise to indicate a score for each of the proposed objectives (column 6 of Table 2.7). The results are produced in Table 2.8.

Table 2.8. Results for the Tooling Industry by Normalization of the Attributes

Attributes	Permanent	Temporary
1) NPV	19,180	3,600
2) IRR	2,000	1,000
3) Payback	2,000	4,000
4) Government	3,000	-
5) Employment	3,000	20,000
6) VA	1,000	2,000
7) Balance of Payments	1,000	250
Totals	31,180	30,850

In considering only one single objective, namely NPV, or in the case of the indifference method, permanent (19,180) would be preferred to temporary tooling (3,600).

If the scores would be considered as weights for the attributes, the same result is obtained. Indeed here, also permanent tooling with 31,180 is preferred to the temporary one (30,850). Weights however do not respond to the statements concerning decreasing marginal utility and nonlinearity.

An approximation responding better to these statements is given by the Reference Point Theory on the basis of a maximal criterion vector (19,180; 2,000; 4,000; 3,000; 20,000; 2,000; 1,000). The answer is found in Table 2.9.

Table 2.9 Deviations from the Reference Point for the Tooling Industry

Attributes	Reference Point	Permanent	Temporary
1) NPV	19,180	0	15,580
2) IRR	2,000	0	1,000
3) Payback	4,000	2,000	0
4) Government	3,000	0	3,000
5) Employment	20,000	17,000	0
6) VA	2,000	1,000	0
7) Balance of Payments	1,000	0	750
Largest deviation		17,000	15,580

Referring to the *Min-Max Metric*:

$$\underset{(j)}{Min} \{ \underset{(i)}{max} (r_i - {}_Nx_{ij}) \}$$

with: $i = 1, 2, ..., n$ as the attributes

$j = 1, 2, ..., m$ as the alternatives

r_i = the ith co-ordinate of the reference point

${}_Nx_{ij}$ = the normalized attribute i of alternative j

it means that temporary tooling is preferred instead of the permanent one. This result is entirely different from the one determined by the other methods, as much importance is given to employment or to the maintenance of employment.

Something similar is possible in the privatization process. We were, however, only considering a period of five years. For project analysis periods from 10 to 30 years are more normal. At that moment it is possible that new industry and privatization would have better repercussions on employment, certainly when the induced or indirect employment is taken into consideration. This means that in any case the results of the long run have to be taken into account as well as for new industry as for privatization.

Reference Point theory would also be very useful for problems of Location Theory, where mostly different objectives and different alternatives are involved.

2.6 PROJECT EVALUATION

The following example of *Pre-Evaluation of Projects* is given. A problem of *Post-Evaluation of Projects* can be solved in the same way.

Suppose the European Commission of the European Union has a budget of 150 million € for small, but very heterogeneous, projects. Per project maximum 75 million € can be spent. The domain can be regional or national.

The following projects did not respond to the constraints:

1) Investment Aid for new firms with more than 20 jobs in new kinds of production 988 m.€

2) Extension of small firms with 1800 new jobs and NPV of 25 m.€ 225 m.€

3) Cleaning-up of industrial sites 82.4m.€

Remaining Projects

1) Assistance in production structure of co-operative farms 69.97m.€
- 40 new co-operative farms
- 400 new young farmers
- jobs per m.€: 400/69.97 = 5.7 jobs

2) Upgrading of tourist infrastructure 67.46 m.€
- employment: 1,000
- jobs per m.€: 1,000/67.46 = 14.8

3) Development of an industrial zone 67.06 m.€
- 300 ha

4) Professional learning with attending school part-time 36.76 m.€
- test with 1,200 students

5) Young unemployed are formed inside the firm 30.38 m.€
- testing 100 trainees

Objectives

Maximization of Employment, trainees, students, hectares
Maximization of Jobs per m.€
Maximization of Quality

Scores by Delphi

A Delphi group of experts is formed, composed half of experts of the Commission and half from outside: private experts and professors in Project Management, Human Resources and General Economics.

Convergence is reached on the following scores:

job: 4; trainees: 2; students: 1; 1ha =: ½ job
jobs per m.€: 100

The Commission finds that quality of the project is still not clearly presented. The experts have to find scores for the general quality of the project. The experts prefer the cardinal numbers: 1,000, 2,000, 3,000 4,000 and 5,000. For each project, the experts reach convergence:

1) 4,000
2) 2,000
3) 5,000
4) 4,000
5) 5,000

Final results

1) jobs: 400 x 4 = 1,600
jobs per m.€: 400/69.97 x 100 = 572

2) jobs: 1,000 x 4 = 4,000
jobs per m.€: 1,000/69.46 x 100 = 1482

3) jobs: 300 x 2 = 600
jobs per m.€: 150/67.06 x 100 = 224

4) jobs: 1,200 x 1 = 1,200
jobs per m.€: 300/36.76 x 100 = 816

5) jobs: 100 x 2 = 200
jobs per m.€: 50/30.38 x 100 = 165

The results are summarized in table 2.10.

Table 2.10 Attributes of Projects for Pre-Evaluation

Attributes	1)	2)	3)	4)	5)
Employment	1600	4000	600	1200	200
Jobs per m.€	572	1482	224	816	165
Quality	4000	2000	5000	4000	5000
TOTALS	6172	7482	5824	6016	5365

If the scores were considered as weights, the *Weighting Method* would produce the following ranking:

Pr.2 P Pr.1 P Pr.4 P Pr.3 P Pr.5

As the budget of the Commission is limited, only the following projects are realized:

1. Project two 67.46 m.€
2. Project one 69.97 m.€

137.43 m.€

The difference, the amount of 12.57 m.€ is not used.

150.00 m.€

Table 2.11 Deviations from the Reference Point with the Maximal Criterion Vector for Pre-Evaluation of Projects

Attributes	Ref. point	1)	2)	3)	4)	5)
Employment	4000	2400	0	3400	2800	3800
Jobs/investment	1482	910	0	1258	666	1317
Quality	5000	1000	3000	0	1000	0
Max. deviation		2400	3000	3400	2800	3800

After the Minimum of the Maximum deviation, the following ranking is obtained:

Pr.1 P Pr.4 P Pr.2 P Pr.3 P Pr.5

This ranking is in disaccord with the weighting method.

As the budget of the Commission is limited, only the following projects are realized:

1. Project one	69.97 m.€
2. Project four	36.76 m.€
	106.73 m.€
The difference, the amount of 43.27 m.€, may still be used for project five	30.38 m.€
	137.11 m.€
Amount, which is not used.	12.89 m.€
	150.00 m.€

Three projects: one, four and five, are realized.

Once again with the *Weighting Method,* there is no room for in-between solutions. On the contrary, a project is chosen for the first place with extreme positions. Indeed, the first ranked, project two, is the best for two of the three objectives. On the contrary, after the *Reference Point Theory*, project two has no chance. Moreover, projects one and four are chosen, which are not excellent in either of the objectives, but on the average for all of them.

NOTES PART 4 CHAPTER 2

[1] A reminder: the Net Present Value represents the difference between the sum of the discounted cash flows during the expected lifetime of the project and the investments. The NPV has to be positive or at least equal to zero. The Break Even Year is the year during the lifetime of the project when the NPV becomes zero.

The Internal Rate of Return is the interest rate that makes the NPV equal to zero.

[2] For more details see Appendix A, Financial Planning in: W.K.Brauers, *Multiple Criteria Decision Making in Industrial Project Management*, Engineering Costs and Production Economics, Elsevier, 1990, 20, 231-240..

[3] For more details, see Appendix A, op.cit. 236-237.

[4] For more details see Appendix B in: W.K.Brauers, *Multiple Criteria Decision Making in Industrial Project Management*, op.cit. 238.

[5] See also Appendix A, Financial Planning in: W.K.Brauers, *Multiple Criteria Decision Making in Industrial Project Management*, op.cit. 236-237

[6] For more details see Appendix C in: W.K.Brauers, *Multiple Criteria Decision Making in Industrial Project Management*, op.cit. 238.

[7] For more details see Appendix A, Financial Planning in: W.K.Brauers, *Multiple Criteria Decision Making in Industrial Project Management*, op.cit. 237.

[8] For more details see Appendix D in: W.K.Brauers, *Multiple Criteria Decision Making in Industrial Project Management*, op.cit. 239-240.

[9] For more details see Appendix E in: W.K.Brauers, *Multiple Criteria Decision Making in Industrial Project Management*, op.cit. 240.

[10] "A large number of the uncounted victims of that slaughter were workers on the avenues around Tiananmen; in the crackdown that followed. Only workers faced execution. Students received jail terms" (Time, *Working Man Blues*, April 15, 2002, 50). The author himself was present in Beijing at that time and remarked that martial law was proclaimed, that the embassies were besieged, that the windows of the hotels were covered with paper against the unwanted regards of tourists etc.

[11] W.K.Brauers, *Scenarios for Updating or Replacement of Old Industry*, the Proceedings of ICHTROI, High Technologies and its Development Strategies, Wuhan, China, 1994, 519-533.

[12.] This was for instance the case with the danger of floods in the Southwest of the Netherlands. The ecologists were firmly opposed to one huge dam closing all Zeeland-islands. The budgeting people were opposed to higher dikes on the islands themselves. Happily the *Scorecard Method* brought a third solution viz., storm-surge barriers (see Part III, 2.3).

[13.] If well-being economics rather than materialistic welfare economics is aimed at i.e., including also social costs (like pollution) and social benefits, the corresponding maximization would be called the maximization of Social Value Added (SVA).

[14] UNIDO, Manual for the Preparation of Industrial Feasibility Studies, UN (ID/206) New York1978.

[15] I.M.D.Little, J.A.Mirrlees, *Project Appraisal and Planning for Developing Countries*, OECD, Heinemann, London, 1974.

[16.] Of course, most of the time the number of different objectives will be less than the numerous objectives enumerated here. This wide choice of objectives is given, in order to strive for completeness and to show objectives with different units of measurement.

CHAPTER 3

REFERENCE POINT THEORY IN SEAPORT PLANNING

The difference between a river port and a seaport is essential for port planning, but some river ports would like to become a seaport. Concerning ports, conflicts may arise between local and national authorities and between the protagonists for a more commercial port against those for a rather more industrial one. Finally, more and more, port planning has to take into consideration ecological conditions.

3.1 COMPARISON OF EFFECTIVENESS OF DIFFERENT CONCEPTS OF SEAPORTS IN AN ANALYTICAL AND PROSPECTIVE WAY, INCLUDING NEW DEVELOPMENTS

First, the purely technical aspect, which is specific for each particular case, is not taken into account. If a port like Wuhan in China would like to become an inland seaport many existing bridges have to disappear, locks to be enlarged and the course of the Changjiang River has to be corrected[1].

A second point to be considered is the study of one single port project or of different projects in competition with one another. The single project will have to pass a filtering stage, but the projects in competition have to be selected through a more complicated method, which also takes optimality into account. This way of thinking is followed here.

Third, two viewpoints are considered: that of a port authority or of a local government and that of the national government. Conflicts between the two standpoints have to be avoided as much as possible. Anyway, a government may consider the seaport as an industrial promotion zone. Nowadays the development of a port is not only seen commercially but also industrially. For a port industrialization means an element of stability in its port traffic. For the industrialist it means better and effective communication with his clients. Consequently, port planning is multiobjective. However, we do not consider the point of view of the private investor who has to choose between a project of industrial or commercial development in a port and a project of industrial or commercial development elsewhere. We are rather involved with the point of view of a port authority and of the national government representing the general interests of the nation.

Fourth, the problem is limited to the installation of a container terminal. Four kinds of seaports are proposed: a seaport near the sea behind locks, a seaport near the sea without locks, an inland seaport on the river itself and an inland seaport behind locks. The Reference Point Theory is used, but without the Maximal Criterion Reference Point. Indeed the characteristics of the several kinds of seaports are so different that an optimum co-ordinate of one of them may not be relevant for another of them. Therefore the stakeholders involved, or their experts, will decide in a Delphi exercise about a *Utopian Criterion Vector*, which seems ideal for them (see Part IV, chapter 1.3).

Finally, a pollution problem is considered in multiobjective analysis of seaports. It concerns a constraint of the salt seawater entering the inland and polluting the drinking water.

3.2 THE FILTERING STAGE IN SEAPORT PLANNING

The Point of View of a Port or of a Local Authority

At this stage, several objectives form a threshold (a lower bound) or a ceiling (an upper limit). In the case of realization of a project from the point of view of a Port or Local Authority it means for instance: a positive Net Present Value, an upper limit on the payback period of Net Present Value, a lower bound for the Internal Rate of Return, a minimum of direct employment creation, a lower bound on hard currency to be provided for investment by foreign sources, a minimum of government subsidies or a maximum on national taxes to be paid, etc. Of course, antecedent to fixation of thresholds and ceilings, the necessary feasibility studies have to be made on market and plant capacity, material inputs, location and site project engineering, manpower and project implementation and evaluation of investment and financing (see Table 3.1).

Table 3.1 The Filtering Stage. Point of View of the Port Authority (20 year period)

Attributes	Units	Thresholds or ceilings
1. <u>Micro-economics</u>		
1.1. Yearly capacity	TEU Units	≥ 500,000 TEU
1.2. NPV (a)	$	≥ 0
1.3. IRR (a)	%	≥ 12%
1.4. Payback period	years, months	Before end of 10th year
2. <u>Macro-economics</u>		
Indirect and direct local employment increase (b)	number of persons per year	≥ 500

(a) Investments are net i. e., after deduction of government subsidies. Revenues include local taxes and rent on industrial land, after government taxes (direct and indirect taxes).
(b) Indirect employment found by I/O tables.

The Point of View of the National Government

National government representing the national interest will stress other points than the port authority or the local government. National government may not allow certain industrial projects to pass the filtering stage, for instance:

- if overproduction of certain products is to be feared;

- if the ratio of debt to equity would exceed a certain proportion.

On the contrary, national government will give special consideration to investment projects, which for instance:

- significantly strengthen the balance of payments, especially through export promotion or import substitution;

- support the country's resource development;

- substantially increase employment;

- conserve energy or replace imported energy supplies;

- establish or develop industries, which form the base for further stages of industrial development;

- support additions to the national wealth;

- use modern machinery and production processes and new equipment except in cases where an older process is to be used if its effectiveness is certified by a reliable institution;

- will be located in a government promotion zone. Such a zone could be for instance at a port. The national government could indeed, create an industrial promotion zone in a port, if further economic development is advisable there.

Suppose the following *Filters* are prescribed by the national government (see Table 3.2)

Table 3.2 . The Filtering Stage. Point of View of the National Government (20 year period)

Attributes	Units	Thresholds or ceilings
1. Well-being economics		
1.1. Degree of salination	mg chloride per l water	10,000 mg/l in subsoil water or in surface water 5 km inland / and 200 mg/l 10 km deep
1.2. Consumer surplus	$	consumer surplus - subsidies given less government. taxes received (all discounted)
1.3. Δ Direct and Indirect employment at the national level	jobs per year	≥ 1,500
2. Macro-economics		
2.1. ENPV (a)	$	≥ 0
2.2. Belonging to a government promotion zone, if existing	yes or no	Yes
2.3. No overproduction due to capacity installed in all the national ports together TEU capacity installed: X_1 =500,000	TEU	1,500,000 TEU ≥ X_1 TEU + X_2 TEU X_2 TEU capacity to be newly installed
2.4. Δ GDP	$	≥ 0
2.5. Surplus in hard currency in the balance of payments	$	≥ 0

(a) ENPV means Economic Net Present Value i. e., discounted revenues before national taxes, minus discounted investments, exclusive of subsidies.

3.3 THE RANKING STAGE IN SEAPORT PLANNING

If the stakeholders are not indifferent, one asks them to attribute scores to each of the objectives after their aspirations. These aspirations have to lead to:

1. an aggregate utility function which is nonlinear;
2. a utility having a well-defined meaning expressed in measurement of effectiveness;
3. learning taking place due to successive Delphi rounds between the stakeholders, determining the importance given to several objectives and to a reference point;
4. stakeholders agreeing with the outcome as they participated in its formulation;
5. an outcome not directed by or in favor of one of the stakeholders in particular but serving the interests of them all.

From the utility per alternative (project) is deducted a ranking of alternatives (projects). However, before considering a ranking, the problem is solved if one alternative dominates all the others for all objectives. The ranking finally takes place only for a set of nondominated alternatives. A contradictory project ranking lower than an already chosen project, with which it is in contradiction, is excluded. A project, which can be repeated several times, is given several rank positions and variations of a project are ranked separately. Finally the first ranked project is chosen first.

Point of View of a Port or of a Local Authority

In the case of choice of industrialization projects by a Port or Local Authority the following objectives could be considered in the ranking stage:

- maximization of Net Present Value (NPV) for the users, expressed in money terms;

- maximization of Internal Rate of Return (IRR) for the users, expressed as an interest percentage;

- minimization of the payback period of NPV for the users, expressed in years and months;

- maximization of NPV for the Port Authority;

- maximization of government subsidies or of similar actions together with minimization of national taxation, expressed in money terms;

- maximization of revenue from rent on industrial land, from local taxes and from other income expressed in money terms;

- maximization of direct and indirect local employment; indirect employment is found through local and national input-output tables while employment is being measured in man-years;

- maximization of hard currency to be provided by foreign sources for investment, expressed in money terms.

Point of View of the National Government

The point of view of the national government can be parallel to the local or port interests. Additionally, however, the national government could consider for the port area:

- maximizing Economic Net Present Value and the increase in Gross Domestic Product;

- maximizing consumer surplus;

- maximizing direct and indirect local and national employment; indirect employment found through local and national input-output tables;

- maximizing hard currency to be provided by foreign sources for investment;

- maximizing the surplus in hard currency on the balance of payments current account.

On the contrary, government objectives could be contradictory to local or port interests. Port interests like to see a maximization of government subsidies and minimization of government taxes and the national government would rather like minimization of local taxes. The Port Authority likes to maximize its NPV, but National Government rather *ENPV (Economic Net Present Value)* for the whole community. At that moment one has to avoid conflicting situations, appealing to Delphi methods for group multiple decision-making. Anyway, the Port Authority or the Local Government may try to have its port recognized by the National Government as a government promotion zone, eventually a condition in a filtering stage.

3.4 APPRAISAL OF PROJECTS FOR THE INSTALLATION OF A CONTAINER TERMINAL IN A SEAPORT OF A DEVELOPING COUNTRY

Competing Projects

A developing country wishes to install a container terminal of at least 500,000 TEU. Five alternatives are proposed.

The first alternative, Project A, consists of the installation at a riverside port, 100 km inland, but on the river itself, capable of receiving the huge container ships. The possibility to bring the huge container ships so far inland is an important advantage of this project, reflected in the willingness of the ship owners to pay high demurrage and local taxes for this solution. It remains however a part of a tidal harbour.

Project B possesses the same advantages as project A belonging also to a riverside port but installed behind locks. This project also means fewer problems with low and high tide, but investments costs are higher given the necessity to dig docks.

Project C is located at a seaport immediately near the sea but behind locks, which means fewer tidal problems, but huge investment costs for digging the necessary docks.

Project D consists of a terminal also immediately near the sea but in open docks i.e., without locks. This means fast delivery of the goods but with a severe problem of salinity, caused by the open dock system at the seaside.

Project E would mean a container terminal on a dam in the sea, meaning fast delivery of the goods. Investment costs however are extremely high translated into high depreciation costs for the dam.

The Filtering Stage

The projects have to respect the constraints, which were already given earlier in this chapter, Tables, 3.1 and 3.2. These tables indicated the thresholds and ceilings dictated by the National Government and by the Port Authority in the filtering process.

Project D is excluded from the side of the National Government, as the degree of saltiness is too high.

Project E is rejected on the base of the prohibitive investment costs, which make NPV as well as ENPV negative.

The Ranking Stage from the Point of View of the National Government

Neither project A, B or C are dominating, which means that a ranking has to bring the solution.

For each project, there is no overproduction given the already installed capacity in container manipulation in all national ports together. Moreover projects A, B and C have no problem with the degree of saltiness. Also the three belong to a government promotion zone. Consequently the three hard constraints have not to be taken into consideration for a further ranking.

Table 3.3 shows the reaction of the projects on the objectives of the national government.

Table 3.3. The Ranking Stage. Reaction on the Proposed Objectives of Projects in Container Terminal (20 year period). Point of View of the National Government

Attributes	Threshold or ceiling	Objectives	Scores to be attributed	Project A	Project B	Project C
1. Social economics						
1.1. Net Δ consumer surplus (a)	≥ 0 $	Max. in $	1 per 1,000$	150 m. $	125 m. $	100 m. $
1.2. Δ dir and ind. employment at the national level (b)	≥ 750 jobs per year	Max. in jobs per y.	100 per job	800 jobs	1,400 jobs	1,500 jobs
2. Macro-economics						
2.1. ENPV (c)	≥ 0 $	Max. in $	1 point per 10,000 $	1 billion $	666 m. $	500 m. $
2.2. Δ GDP	≥ 0 $	Max. in $	1 point per 10,000 $	2.5 billion $	2.6 billion $	2.7 billion $
2.3. Surplus in hard currency in the Bal. of Payments	≥ 0 $	Max. in $	1 point per 10,000 $	1,200 m. $	900 m. $	700 m. $

(a) Net consumer surplus = consumer surplus - subsidies given less government taxes received (all discounted).
(b) Indirect employment found with the I/O tables.
(c) ENPV means Economic Net Present Value i. e., discounted revenues before national taxes, minus discounted investments, exclusive of subsidies.

The objectives are measured by attributes with different units, namely dollars and jobs. However, even the significance of a dollar is different from objective to objective (see column 4 of Table 3.3).

In order to normalize the attributes in a non-subjective way, a Delphi group of stakeholders around the national government, or of their experts, attributed scores to each attribute (column 4 of Table 3.3). On basis of Table 3.3 the number of scores per project can be calculated.

After the Delphi normalization of the objectives the stakeholders have to determine a reference point. Several kinds of reference points are possible, as was already demonstrated earlier. The Maximal Criterion Point is the most realistic, as the co-ordinates are deducted from the proposed alternatives themselves. However, the proposed projects on seaports are so heterogeneous that the use of their co-ordinates is not realistic. Rather the stakeholders have to fix a reference point, which after their opinion, is the *Ideal Point*. Once again as a tribute to objectivity the Delphi group has to meet.

Suppose the stakeholders around the Government agree upon the following Reference Point:

net increase in consumer surplus	200 m. $
increase in employment	1,500 jobs
ENPV	1 billion $
increase GDP	2.7 billion $
surplus hard currency	1.5 billion $

Table 3.4 calculates the deviations from the Reference Point, which is here an Ideal Point.

Table 3.4 Deviations from the Reference Point per Project for a Container Terminal (20 Year Period). Point of View of the National Government

Attributes	Ref. Point	Project A	Project B	Project C
1. Social Economics				
1.1. Net ΔConsumer Surplus	200,000	150,000	125,000	100,000
1.2. Δ Dir. and Indirect Employment at the National Level	150,000	80,000	140,000	150,000
2. Macro-Economics				
2.1. ENPV	100,000	100,000	66,600	50,000
2.2. Δ GDP	270,000	250,000	260,000	270,000
2.3. Surplus in Hard Currency in the Bal. of Payments	150,000	120,000.	90,000	70,000
	Largest Deviation	70,000	75,000	100,000

At that moment the ranking is as follows:

1) Project A, namely a container terminal at a riverside port on the river itself;
2) Project B, viz., a container terminal at riverside port behind locks;
3) Project C, viz., a container terminal near the sea behind locks.

Ranking Stage from the Point of View of the Local Government or of the Port Authority

Table 3.5 indicates the reaction of the projects on the objectives of the local government or of the port authority.

Table 3.5 The Ranking Stage. Reaction on the Proposed Objectives of Competing Projects for Container Terminal (20 year period). Point of View of the Port Authority

Attributes	Threshold or ceiling	Objectives	Scores to be attributed	Project A	Project B	Project C
1. Micro-economics						
1.1. Payback period	before end of 10th year	Minimum in years, months	1,000 per month earlier	3 years, 1 month	6 years, 2 months	6 years, 3 months
1.2. NPV (a)	≥ 0	Max. in $	1 per 10,000 $	1,180 m. $	866 m. $	680 m. $
2. Macro-economics						
Increase indirect and direct local employment (b)	$\geq$ 500 jobs per year	Max. in jobs per year	100 per job	500 jobs	700 jobs	750 jobs

(a) Net Present Value. Revenues are incl. local taxes, rent on industrial land, but after government taxes (direct and indirect taxes). Investments are net i. e., after deduction of government subsidies.

(b) Indirect local employment found with input-output tables.

The stakeholders around the local government or of the port authority normalized the objectives in a Delphi kind of operation, as shown in column 4 of Table 3.5. In this way, the number of scores per project can be calculated.

Next, the stakeholders have to determine a reference point. Several kinds of reference points are possible, as was already demonstrated earlier. The Maximal Criterion Point is the most realistic one, but the stakeholders have to fix rather a reference point, which after their opinion, is the ideal point. Once again as a tribute to objectivity the Delphi group has to meet.

Suppose the stakeholders around the local government agree upon the following Reference Point:

payback period	6 years
NPV	1,200 m. $
increase employment	750 jobs

It is understandable that stakeholders around a government consider a payback period as less important. Payback period belongs rather to financial economics, whereby shareholders, eager for the regularity of their profits, are more interested in it. Table 3.6 calculates the deviations from the Reference Point

Table 3.6 The Ranking Stage. Deviations from the Reference Point for Projects in Container Terminal (20 year period). Point of View of the Port Authority

Attributes	Reference Point	Project A	Project B	Project C
1. Micro-economics 1.1.Payback period	48,000	83,000	47,000	46,000
1.2. NPV	120,000	118,000	86,600	68,000
2. Macro-economics Increase indirect and direct local employment	75,000	50,000	70,000	75,000
	Largest deviation	35,000	33,400	52,000

At that moment, the ranking is as follows:

1) Project B
2) Project A
3) Project C

Unhappily, there is a conflicting situation between the attitudes of the Government and the Port Authority. The National Government may apply its dominating position towards the local authority or port authority.

Otherwise in the case of a far-reaching autonomy of the local government or of the port authority, an open discussion between the two groups of stakeholders is needed. It is, however better to use a nominal method such as the Delphi kind of exercise between the two partners (the way to proceed is explained in Part III, 1.5).

NOTES PART 4 CHAPTER 3

[1] A paper on this subject was presented by the author at the International Symposium on Development Strategies of Riverside Cities in Wuhan on October 16-20, 1991.

CHAPTER 4

PRIVATIZATION AND REFERENCE POINT THEORY

Nowadays a general tendency in the world exists towards privatization of public enterprises and services. Privatization means then that government services or state enterprises are turned over to private ownership. The purpose is mainly to create a better management for these institutions. However, the ultimate goal is rather vague as not only the government and the eventual buyer, but also the employees of these institutions and even the whole population, given the repercussions on the general tax level, are interested in the issue. All these stakeholders may pursue different and even contradictory objectives.

4.1 THE PROBLEM[1]

For the government the employment goal has to be an independent objective separated from an optimal takeover price. This price could include a compensation for unemployment, but the unemployed may not be satisfied with a monetary compensation alone. The government may also ask for pollution abatement and for new investments as most of these public enterprises show a shortage in investment. In fact, the government plays a dual role. On the one side, it is the defender of the general interest. On the other side it takes care of its own interests, namely to obtain a maximum of takeover price in case of privatization.

As said before, many stakeholders will try to influence decisions. Among the decision influencing agents, pressure groups may lobby for their interests such as trade unions, ecologists and consumer organizations. This means that in fact *Social Welfare* and even *Social Well-Being*, think of pollution, has to be optimized given multiple independent objectives.

Most of the time with privatization multiple objectives are faced, such as selling price, employment, new investments and pollution abatement. The optimization of these objectives is then judged case by case, sometimes in a rather subjective way, namely by feeling (Fingerspitzengefühl). Consequently, there is a need for a more general and objective method, that is to say a scientific method, which can compare several takeover bids for privatization of a public service or a firm optimizing multiple objectives, sometimes with different units of measurement.

Additionally from what is concluded above, stress is laid upon convergence, including eventually the opinions of minority groups and on optimality to the greatest extent possible.

4.2 THE FILTERING STAGE IN THE CASE OF PRIVATIZATION

At this stage, several *hard constraints* are determined as a threshold (a lower bound) or as a ceiling (an upper limit). In the case of privatization, the filter can be either of a qualitative or of a quantitative nature. As an example of a qualitative limit, let us give the example of the takeover of a public enterprise by a daughter of a multinational. In that case the multinational has to accept responsibility for the possible bankruptcy of the daughter, let us say for the next ten years. Indeed out of past experience it is known that bankruptcy of a daughter firm of a multinational is the cheapest way to liquidate that firm. Another example is the obligation to invest in a well-defined region. Quantitative limits could be for instance: a minimum bottom price, a minimum of guaranteed employment, a minimum of new investment, maxima for pollution emissions (CO_2, CO, NO_x, SO_2, particulates, etc.), the liquidity position of the applicant, the obligation to accept the government as a majority or as a minority shareholder, the consideration of a risk factor, etc.

In Part IV, Chapter 2 of this book, it was shown that, antecedent to the fixation of thresholds and ceilings, the necessary feasibility studies have to be made on market capacity, material inputs and human resources and on the evaluation of the necessary investment and financing. In addition, in the presence of a single takeover proposition for a public enterprise or service, the model stops after the filtering stage.

Sometimes a private partner discusses in advance the privatization with the public enterprise or service to be privatized. In this way, many difficulties are already solved before the offer is officially introduced. Nevertheless, the government or its representatives have to decide if a public auction is not better for the general interest.

With several takeover bids, which already passed the filtering stage, the next step will be that they have to face either the indifference or the ranking stage. Let us discuss first the indifference stage.

4.3 AN INDIFFERENCE STAGE IN PRIVATIZATION

In real life, many stakeholders are indifferent about the alternative solutions, if some thresholds or ceilings are respected. Most of the commodities, however, are not free commodities and services and in the traditional scarce economy, the cost side takes then a decisive position. Therefore, in an indifference stage, when the alternatives pass the filtering stage and the stakeholders are indifferent about the alternatives, the cost side will determine a ranking of the possible alternatives[2].

In the case that government decision makers are indifferent if thresholds are respected by the several propositions concerning employment, investments, liquidity etc., at that moment, the takeover bid alone will be decisive for the ranking of propositions with a maximum possible price for the government.

In the 1991 drive for privatization, the Brazilian government was indifferent about new investments in and the effective way of working of the new private firm and about the decrease in employment. Indeed, it was assumed that a private firm will automatically provide new investments, with a multiplier effect on the national income and on the government income through taxation. The decrease in employment was not considered as important, because in 1991 ten years of restrictions on hiring new personnel already existed for the public sector. The firing of 10 to 15% of personnel from the firms being privatized was considered as normal. In this way for the Brazilian government the selling price of the firm remained then the only objective to be maximized. The bottom price for the public auction was then the economic value of capital for production i.e., the sum of the discounted expected cash flows as well known from project analysis[3]. Two separate consulting firms estimated the economic value of capital for production of a firm. In case of strong differences in the estimation, a third consulting firm was approached.

After the filtering stage and when the indifference stage is not applicable, the remaining propositions for privatization have to be ranked for the several objectives considered simultaneously.

4.4 SIMULATION EXERCISE FOR PRIVATIZATION IN UKRAINE

The Problem

At the occasion of an international workshop in Ukraine in 1992, the intention was to come to a report on privatization in Ukraine[4]. At that moment, however, there were not enough data available, the banks were not organized, there was no official money and credit cards were not accepted.

Therefore, the following simulation was proposed based on semi-fictive data for a public enterprise.

The government or its proxies are supposed to fix for instance the following objectives:

a) in the micro-economic field:
   1. maximization of the selling price for the public enterprise or service;
   2. maximization of guaranteed employment;
   3. maximization of new investments over a ten year period;
   4. minimization of the payback period of Net Present Value[5];

b) in the macro-economic field:
   1. maximization of discounted Value Added in constant prices over a ten year period;
   2. maximization of total employment (direct, indirect and secondary);

3. maximization of discounted net exports in constant prices over a ten year period;

c) in the social-economic field:

- minimization of the emission of sulfur dioxide content taken as a typical example of pollution.

The Filtering Stage of the Simulation

Suppose the government or its proxies fix the following thresholds or ceilings for eventual takeover bids on public enterprises or services besides qualitative requirements such as qualitative filters on location, financial responsibility of multinationals for their daughters, etc.:

- a minimum bottom price for the takeover bid;
- a minimum of guaranteed employment;
- a minimum of new investments over a ten year period;
- a maximum on the emission of sulfur dioxide;
- a minimum of Internal Rate of Return e.g., the rate for getting a loan from the World Bank;
- a maximum on the payback period of Net Present Value.

Suppose that seven takeover bids are made which will be called A, B, C, D, E, F and G. Takeover bid D does not guarantee the threshold and ceiling concerning respectively Internal Rate of Return and payback period. Takeover bid E cannot guarantee the employment level and refuses to invest in the indicated region, F refuses new investments and G does not accept any anti-pollution rule and the financial responsibility of the mother firm. This means that after the filtering stage takeover bids A, B and C remain. The reference point theory will be applied on these takeover bids.

The Ranking Stage of the Simulation

After the available information none of the remaining takeover bids seems to dominate strongly or weakly the others for all objectives (see next Table 4.1). The table also shows the different units of measurement of the objectives.

In order to normalize the objectives the decision makers are asked in a Delphi exercise to indicate a non-additive score for each of the proposed objectives, which are expressed in different units of measurement.

The result of the score exercise is given in column (7) of Table 4.1.

Table 4.1 Reaction of the Several Takeover Bids on the Proposed Objectives

1.Attributes	2. Filtering	3. Objectives	4. A	5. B	6. C	7. Non-additive scores
Entrepreneurial Economics						
1. Takeover bid	≥ 0.8 bln $	max. in $	0.8 bln $	1 bln $	1.5bln.$	1 per 100,000 $
2. Guaranteed employment	≥ 750 jobs	max. jobs	750 jobs	800 jobs	900 jobs	10 per job
3. Payback period	≤ 9 years	min. in years	5 years	7 years	9 years.	100 per year earlier
4. New investments	≥ 2.5 bln $ over 10 years	max. in $	4.5 bln $	3 bln.$	2.5 bln. $	1 per 100,000 $
5. Internal Rate of Return	≥ 13% over 10 years	-	yes	yes	yes	yes
Macro-economics						
5. Total employment	-	max. jobs	1,300 jobs	1,200 jobs.	1,100 jobs	5 per job
6. V.A. (discounted)	≥ 0 over 10 years	max. in $	800 m. $	600 m. $	400 m.$	1 per 80,000 $
7. Exports (discounted)	≥ 350 m.$ over 10 years	max in $	450 m. $	400 m. $	350 m. $	1 per 100,000 $
8. Risk factor: deterioration of Terms of Trade	≤ 50% of exports	min. in %	30%	50%	20%	100 per % point less
Social Economics						
9. Emission of sulfur dioxide	≤ 8 ton per year	min. in tons	8 tons	7 tons	5 tons	1,000 per ton less
Qualitative Attributes						
1. Regional	in region I	-	yes	yes	yes	yes
2. Financial Responsibility	Yes	-	yes	yes	yes	yes

In Table 4.2 the deviations of the co-ordinates are given after the Maximal Criterion Values.

Table 4.2 Deviations of the Co-ordinates of the Takeover Bids from the Co-ordinates of the Reference Point

Attributes	Reference Point	A	B	C
1. Takeover bid	15,000	7,000	5,000	0
2. Guaranteed employment	9,000	1,500	1,000	0
3. Payback period	400	0	200	400
4. New investments	4,500	0	1,500	2,000
5. Total employment	6,500	0	500	1,000
6. VA	10,000	0	2,500	5,000
7. Exports	4,500	0	500	1,000
8. Risk factor	3,000	1,000	3,000	0
9. Pollution	3,000	3,000	2,000	0
Largest deviation		7,000	6,000	5,000

The ranking is made after the smallest size of the deviations (see Part IV, 1.5). At that moment the ranking for the best choice is as follows:

1) takeover bid C
2) takeover bid B
3) takeover bid A.

However, the differences are not very large between the three propositions.

Is a reverse order in ranking possible if a new proposition is introduced? Therefore we return to the main example as given in the previous tables. Suppose a new takeover bid M with the same co-ordinates as C, with exception of a total employment guaranty of 12,000. Table 4.2 changes then as follows:

Table 4.3 Changes in Table 4.2 with another Maximal Criterion Value for Employment

	Reference Point	A	B	C	M
5) Total employment	12,000	5,500	6,000	6,500	0
Largest deviation		7,000	6,000	6,500	5,000

This table produces the following ranking:

M P B P C P A instead of: C P B P A

The reason of this reverse order is simple: the reference point is changed. The following co-ordinates characterized indeed the first reference point:

(15,000; 9,000; 400; 4,500; *6,500*; 10,000; 4,500; 3,000; 3,000),

but the second by: (15,000; 9,000; 400; 4,500; *12,000*; 10,000; 4,500; 3,000; 3,000).

If we do not change the co-ordinates of the reference point, then there is no reverse order in ranking possible (see Table 4.4)[6].

Table 4.4 Changes in Table 4.2 with a new Proposition but with the Original Maximal Criterion Value for Employment

	Reference Point	A	B	C	M
5) Total employment	6,500	0	500	1,000	5,500
Largest deviation		7,000	6,000	5,000	5,500

Result: C P M P B P A.

Takeover bid B takes a midway position in the sense that B is never a champion in one of the objectives (see Tables 4.2, 4.3 and 4.4). Nevertheless, its position is honorable, which would not be the case with a weighting method.

4.5 SOME PRACTICAL CONSEQUENCES OF APPLICATION OF REFERENCE POINT THEORY FOR PRIVATIZATION

The rigidity in application is one of the objections which could be raised against the application of Reference Point Theory for privatization. In order to answer this remark the several phases of the model will be reconsidered.

(1) The government has to fix its objectives. This phase has always to be present though many governments forget to state clearly their objectives.

(2) The private proposals for takeover have to be filed, which is also necessary for privatization without a model.

(3) The distance function used is a very simple function for which the software is very easy to write even for a personal computer. Output will also be immediately evaluated when a new proposition is introduced.

In this way, it is shown that the model is very flexible with exception perhaps of the Delphi exercise, in fact a one-off operation. It would be nice if the simulation presented here could be tested better.

In a previously Centrally Planned Economy like Ukraine, the state enterprises were in bad shape with very poor management, low productivity and an urgent need for new investments and working capital, being not immediately ready to face competition in a free market economy.

Many preliminary conditions towards a market economy had to be fulfilled before Ukraine could think of a privatization based on multiple objectives of Social Well-Being[7].

The Reference Point Theory model was also presented to *Treuhand*, the official body responsible for privatization in former Eastern Germany. A member of the board of this important body stated that privatization was mainly done on pragmatic and political grounds. In fact, Treuhand wished fast solutions for its privatization proposals. However, not much enthusiasm was shown for the takeover of the nationalized enterprises. Consequently, in 1994 Treuhand stopped its activities with an enormous deficit, instead of the expected surplus in takeover returns. The unemployment due to privatization was underestimated, which created a huge wave of unemployment and consequently much social unrest in former Eastern Germany. Instead of a mono-objective approach, multi-objectivity would have maintained the employment level in the short and medium term with new investments to increase labor productivity on the long run.

If a model for privatization has to be developed it has to take into consideration *Social Welfare* and even *Social Well-Being*. At that moment, many stakeholders are interested in the issue. They will postulate many different objectives, sometimes expressed in different units of measurement.

It is true, however, that the economic axiom is accepted whereby all objectives are measurable. Nevertheless, the different units of measurement cannot be reduced to a single unit. At that moment traditional methods such as cost-benefit, cost-effectiveness, weighting and voting in order to satisfy "as much as possible all stakeholders and all objectives" cannot be applied. On the contrary, Reference Point Theory, first with a filtering stage, then with a ranking stage and finally with the use of a distance function, will bring a univocal ranking of all the privatization proposals. The model is flexible enough and with adequate computerization, to deal with the peculiar problems of privatization.

NOTES PART 4 CHAPTER 4

[1] A first version of this text was published as: W.K.Brauers, *Privata: a model for privatization with multiple non-transitive objectives*, Public Choice, Kluwer Academic Publishers, 1995, 85, 353-370.

[2] See therefore Part III Chapter 4 of this book.

[3] See therefore Part IV, Chapter 2 of this book. For the definition of "the economic value of capital for production", and for similar notions of capital value, see also: W.K.Brauers, *Measurement of Capital for Production in Macro-Economics*, Institute for developing Countries, RUCA, University of Antwerp, 1990.

[4] IIASA Workshop on Intelligent Decision Support System, Academy of Sciences, Kutzively, Crimea, Ukraine, October 1-8, 1992.

[5] The payback period was included as a control on the seriousness of the takeover bid viz., concerning employment, new investments and value added.

[6] It has to be remarked that only a reverse order in ranking created outside the decision maker's opinion has to be condemned. The decision maker himself may however change his priorities with a reverse order in ranking as a result.

[7] Anyway these conditions were not fulfilled in 1992. See: W.K.Brauers, *Privatization with Several Objectives. The Case of Ukrainian Privatization*, Working Paper, Faculty of Applied Economics, University of Antwerp (RUCA), 1992. Since then important decisions for privatization were taken in Ukraine (Wall Street Journal Europe, March 23,1999,1). The transition remains, however, difficult: M.Costello, *First find your Ukrainian*, Strategy, the bi-monthly magazine of the Strategic Planning Society, London, May 1998, 12-14.

CHAPTER 5

MARKETING WITH MULTIPLE OBJECTIVES FOR CONSUMER GOODS WITH APPLICATIONS FOR A DEPARTMENT STORE

A department store has a number of objectives. How is it possible to strive for the optimal, or as much as possible optimal, fulfillment of all these objectives simultaneously? The use of weights leads to extreme results with few chances for an in-between alternative. Therefore, the Reference Point Theory is applied.

5.1 OBJECTIVES FOR A DEPARTMENT STORE

The several objectives of a department store concern the turnover, either with deepening of the home market or with penetration abroad; further, the profitability and productivity aspects, the choice of the best location, government support by increasing the employment level, by Value Added and eventually by a positive influence on the Balance of Payments. In addition, the support is needed of the personnel, the trade unions, the shareholders and the clients. These stakeholders like to enjoy an optimum position too.

In order to define an objective better we have to focus on the notion of *Attribute*. For instance, if generating of new employment is the objective, the attribute could be: guaranteeing at least 1,000 new jobs over a certain period. An attribute should always be measurable.

Assume a number of mutually competitive alternatives, that we call here *Projects,* faces the attributes and objectives. For instance, a simulation for a large department store takes into account four projects, called A, B, C and D.

5.2 FILTERING AND NORMALIZATION IN THE CASE OF A DEPARTMENT STORE

Filtering Stage

Suppose a large department store decided to use the model to evaluate the four projects of implantation of supermarkets either national or international.

Some projects did not pass the filtering stage (see Table 5.1).

Table 5.1 Filtering for a Department Store

Attributes	Thresholds or Ceilings
Entrepreneurial economics	
1. NPV	≥ 0 $
2. IRR	≥ 12%
3. Payback Period	before end of the 5th year
4. Taxes min. Subsidies (incl. para-fiscal)	≤ 30% of VA
5. Penetration Domestic	s.q.
6. Penetration International	
6.1. Other Industrialized Country	s.q.
6.2. Country Southern Europe	s.q.
6.3. Country Eastern Europe	s.q.
7. Deterioration Terms of Trade possible	≤ 50% NPV
Macro-economics	
1. Direct Employment (national)	≥ 500 permanent jobs
2. Indirect + Secondary Employment (national)	≥ permanent 250 jobs
3. Gross VA	≥ 1 billion $
4. Positive influence Balance of Payments	≥ 1 million $

The stakeholders decided for a hard constraint on the Internal Rate of Return of 12% and Project D did not reach that rate. On the contrary, projects A, B and C passed all the constraints.

Project A expands less in the domestic market, but rather in one country of Southern Europe and in one country of Eastern Europe. This expansion means a higher risk ratio, mainly because import prices for these countries will rise more than domestic and export prices. Project C is more interested in the domestic market. Project B expands in the domestic market and in another large industrialized country, but is in fact mainly situated between the two other projects. In other words A and C take more extreme positions than B.

After the available information none of the remaining projects seems to dominate the others for all objectives simultaneously. A solution has to be found for projects A, B and C forming a *Set of Until Now Incomparable Alternatives*. Moreover, the attributes of the objectives are expressed in different units of measurement.

Normalization

In order to normalize the attributes, the stakeholders are asked in a nominal exercise about the importance of each of the proposed objectives. The nominal exercise is in fact a Delphi exercise in the following way:

- a group of stakeholders or their delegates, for instance experts, represents all the stakeholders;
- their computer assisted meeting is organized in such a way that communication between them is directly impossible (anonymous). Anonymity could also be guaranteed if the meeting is replaced by sending questionnaires;
- the following directives are given after all objectives and attributes are clearly shown:
  - ask the question: "do you suggest other important objectives besides the objectives found after a brain storming session of a steering group"?
  - for each attribute a score is given by each participant;
  - the medians and quartiles for each score are calculated;
  - if the deviations are too large, explanations are asked about the extreme positions and a new round is organized;
  - a third round will give an outcome for the scores, but new rounds may be organized until convergence in opinion is satisfactory and after the steering group or the computer has indicated some anomalies to be solved (Table 5.2).

Table 5.2 Reactions of Competing Projects in Supermarket Structure on the Proposed Objectives (10 Years)

Attributes	Objectives	Project A	Project B	Project C	Scores
<u>Entrepreneurial economics</u>					
1. NPV	MAX.	2,000 m. $	1,000 m. $	800 m. $	5 per m. $
2. IRR	MAX.	17%	14%	13%	1,000 per interest point more than 12%
3. Payback Period	MIN.	before end of 5th year	before end of 5th year	before end of 4th year	1,000 per year earlier
4. Taxes min. Subsidies (incl. para-fiscal)	MIN.	20%	25%	30%	200 per % point less than 30% VA
5. Penetration Domestic	MAX.	+0.5%	+4%	+10%	500 per % point
6. Penetration International					
6.1. Other Industrialized Country	MAX.	+2%	+1%	-	500 per % point
6.2. Country Southern Europe	MAX.	+20%	+15%	-	300 per % point
6.3. Country Eastern Europe	MAX.	+49%	+30%	-	100 per % point
7. Deterioration Terms of Trade possible	MIN.	NPV	10% NPV	0% NPV	200 per % point less than 50%
<u>Macro-economics</u>					
1. Direct Employment (national)	MAX.	500 jobs	1,100 jobs	2,000 jobs	1 point per job
2. Indirect + Secondary Employment (national)	MAX.	250 jobs	550 jobs	1,000 jobs	1/2 point per job
3. Gross VA	MAX.	5.5 billion $	11 billion $	12.5 billion $	200 per billion $
4. Positive influence Balance of Payments	MAX.	5 m. $	4 m. $	3 m. $	100 per m. $

5.3 THE APPLICATION OF THE REFERENCE POINT THEORY WITH MAXIMAL CRITERION VALUES FOR A DEPARTMENT STORE

The co-ordinates are equal to the highest number of points per attribute found in Table 5.3 (Maximal Criterion Values).

Table 5.3 Number of Points obtained per Project and Co-Ordinates of the Reference Point equal to the Maximal Criterion Values (10 Years)

Attributes	Project A	Project B	Project C	Ref. point
1. NPV	10,000	5,000	4,000	10,000
2. IRR	5,000	2,000	1,000	5,000
3. Payback Period	-----	-----	1,000	1,000
4. Taxes	2,000	1,000	----	2,000
5. Penetration Domestic	250	2,000	5,000	5,000
6. Penetration. Industr. Country	1,000	500	-	1,000
7. Penetration Southern Europe	6,000	4,500	-	6,000
8. Penetration Eastern Europe	4,900	3,000	-	4,900
9. Deterioration Terms. of Trade.	2,000	8,000	10,000	10,000
10. Direct Employment	500	1,100	2,000	2,000
11. Indirect + Secondary Employment	125	375	500	500
12. Gross VA	1,100	2,200	2,500	2,500
13. Influence on Balance of Payments	500	400	300	500
Totals	33,375	30,075	26,300	50,400

In case of linearity and when the scores are translated into corresponding weights, the totals of Table 5.3 would have a meaning. The ranking of the weighting concept would then be:

$$A P B P C$$

Due to the law of diminishing marginal utility, nonlinearity is the rule. Conformably to these requirements, the reference point theory is used with the maximal criterion values as the co-ordinates for the reference point. This means that the corresponding highest co-ordinates of the projects are assigned to the reference point as shown in Table 5.3.

In Table 5.4, the deviations are given of the co-ordinates.

Table 5.4 Deviations of the Projects Co-Ordinates from the Co-Ordinates of the Reference Point (10 Years)

Attributes	Reference Point	Project A	Project B	Project C
1. NPV	10,000	0	5,000	6,000
2. IRR	5,000	0	3,000	4,000
3. Payback Period	1,000	1,000	1,000	0
4. Taxes	2,000	0	1,000	2,000
5. Penetration Domestic	5,000	4,750	3,000	0
6. Pen. Industr. Country	1,000	0	500	1,000
7. Pen. Southern Europe	6,000	0	1,500	6,000
8. Pen. Eastern Europe	4,900	0	1,900	4,900
9. Deterioration T. of T.	10,000	8,000	2,000	0
10. Direct Employment	2,000	1,500	900	0
11. Ind. + Second. Empl.	500	375	125	0
12. Gross VA	2,500	1,400	300	0
13. Balance of Payments	500	0	100	200
Largest deviation		8,000	5,000	6,000

At that moment, a minimization of the largest deviation has to take place. The ranking for the best choice is then as follows:

$$B\,P\,C\,P\,A$$

1) Project B, expands in the domestic market and abroad. It brings a midway solution between A and C with no extreme positions of its responses to the attributes;
2) Project C, expands alone in the domestic market;
3) Project A, expands in the domestic market, but mainly in foreign countries.

This ranking contrasts with the linear case, at the moment that the weights take the same proportions as those of the scores.

5.4 A CHANCE FOR AN IN-BETWEEN ALTERNATIVE

Linearity with weights favor extreme positions. Reference Point Theory may give a chance for an in-between alternative, such as alternative B in the marketing example.

The fact that Reference Point Theory takes into account an in-between alternative may have a boomerang effect. Indeed Reference Point Theory may give support to the *Egalitarian Principle*, as stated by Myerson[1]:

> "The egalitarian principle in ethical theory asserts that the best social policy is the one which gives the greatest welfare subject to the constraint that all individual members should enjoy equal benefits from society".

Let us take the example of Myerson but this time translated in a Reference Point Theory approach.

A father is limited in his budget to finance the studies of his twin sons.

A Ceiling as a Filter

Assume costs as follows: "teacher" studies cost double that of "clerk" studies (e.g., 2 against 1); "doctor of medicine" studies cost three times more than for "teacher" studies (e.g., 6 against 2). The budget of the father is limited to seven points. This ceiling excludes two studies in medicine. Three alternatives remain: one son a doctor of medicine (the brightest) the other a clerk, or either becoming teachers or both clerks.

Application of the Maximal Criterion Value

The Maximal Criterion Value produces the Reference Point (6,6).

Table 5.5 Reference Point Theory with Maximum Criterion Value and Cost of Studies Deviations

Reference Point (6,6)	Doctor/Clerk	Teachers	Clerks
Education First Son (MAX.)	6 – 6 = 0	6 – 2 = 4	6 – 1 = 5
Education Second Son (MAX.)	6 – 1 = 5	6 – 2 = 4	6 – 1 = 5
Largest deviation	5	4	5

Smallest-largest deviation means that preference is given to teachers instead of to doctor/clerk studies, i.e., an *Egalitarian Result*. However, all depends on the way the objectives are decided.

As Myerson remarks there could be a coin tossing between the sons. Each son would prefer a .50-.50 gamble between becoming a doctor or a clerk to becoming a teacher. At that moment, the result is no longer egalitarian, but Pareto optimal as the whole budget of the father, namely seven, is used as shown in the following Table 5.6.

Table 5.6 Maximum Criterion Value and Cost of Studies Deviations with a Randomization Plan

Reference Point (6,6)	Doctor/Clerk	Teachers	Clerks
Education First Son (MAX.)	(6 – 6) * 0.5 = 0 (6 – 1) * 0.5 = 2.5	6 – 2 = 4	6 – 1 = 5
Education Second Son (MAX.)	6 – 6 = 0 (6 – 1) * 0.5 = 2.5	6 – 2 = 4	6 – 1 = 5
Largest deviation	2.5	4	5

The doctor/clerk scheme is followed and the father allocates six to the lucky son and one to the unlucky son, letting a coin toss decide, i.e. an *Utilitarian Result*.

5.5 CONCLUSIONS ABOUT REFERENCE POINT THEORY IN CONNECTION WITH THE EXPANSION OF A DEPARTMENT STORE

After filtering and domination, incomparable alternatives could remain for the expansion of a department store. Reference Point Theory is proposed as a solution. Therefore, distances are measured from the alternatives to a reference point. This reference point could be realistic or idealistic. Realistic means that the co-ordinates of the reference point are deduced from co-ordinates of the alternatives (Maximal Criterion Value). On the contrary, an ideal point is defined as an ideal situation for all stakeholders concerned. Therefore, in order to be as non-subjective as possible, the representatives of the stakeholders would gather, for instance, for a Delphi exercise.

The *Ideal Point* was chosen for marriage candidates as an approach for performance. *Performance Management* is important for students' evaluation, but also for any performance in the public or in the private sector. In the case of seaport planning, the projects are so divers, that the use of their co-ordinates is not realistic. Therefore, preference is also given to an ideal point.

The *Maximal Criterion Value* was applied in connection with the expansion of a department store, for Project Analysis, for Evaluation of Projects, and for Privatization.

NOTE PART 4 CHAPTER 5

[1] R.B.Myerson, *Utilitarianism, Egalitarianism, and the Timing Effect in Social Choice Problems*, Econometrica, Journal of the Econometric Society, vol.49, N°4, 1981, 883-897.

PART FIVE

A MULTIPLICATIVE FORM FOR MULTIOBJECTIVE OPTIMIZATION

CHAPTER 1

THE MULTIPLICATIVE REPRESENTATION OF MULTIOBJECTIVE OPTIMIZATION

Human behavior is characterized by pursuing several objectives simultaneously. These objectives are sometimes contradictory, for instance obtainment of richness and economic growth as opposed to diminution of pollution and other monetary or social costs. Such objectives could even be more immaterial such as happiness and the absence of sorrow.

One might even go further in seeking optimal fulfillment of all objectives at the same time or attainment of this optimum as much as possible. An *Objective* indicates the direction in which we should strive. In order to better define an objective we must refer to the notion of *Attribute*. An attribute measures the objective. This measurability is either ordinal (ranking) or cardinal (a collection). In the cardinal approach, attributes are expressed in physical units, for instance using metric system units or money terms. Other indicators may also characterize attributes. Objectives seek solutions from a set of alternative potential actions, policies or projects. We refer to these simply as *Alternatives*.

Is it possible to bring all attributes and objectives together in one utility function?

1.1 FORMS OF THE UTILITY FUNCTION

The decomposable utility function over n attributes of the objectives:

$$f[u_1(x_1), u_2(x_2), \ldots, u_i(x_i), \ldots, u_n(x_n)]$$

is best known under its additive form with the use of weights:

$$u(x_j) = \Sigma_i\, w_i \,.\, x_{ij} \qquad \text{(I)}$$

where: $u(x_j)$ = total utility for the multiple objectives of alternative j

$j = 1,2,\ldots,m$; m the number of alternatives

w_i = weight for attribute i where $\Sigma\, w_i = 1$

$i = 1,2,\ldots,n$; n the number of attributes

x_{ij} = response of alternative j on attribute i

Human behavior is also characterized by a function incorporating decreasing marginal utility. Therefore, linearity with the use of weights as a multiobjective approach has to be avoided.

Linearity and the use of weights can lead to extreme results with few chances for an alternative situated between these extremes. Using weights an in-between point like C(30,60) could never rank first in comparison to the points A(10,100) and

B(55,10), though the three points belong to the same straight line $2x + y = 120$. On the contrary taking into consideration decreasing marginal utility and indifference curve analysis, i.e., nonlinearity, it is quite possible that point C dominates both points A and B. If linearity is rejected, also piecewise linearity has to be rejected too. Indeed, piecewise linearity, as well as linearity, uses weights with all the shortcomings.

In addition, generalization after a two by two comparison of objectives can lead to contradictory results or to reverse orders, as was demonstrated earlier.

It was thought that *Reference Point Theory* could bring a solution, as was described in Part IV of this book. In any case, with reference point theory normalization has to occur. Normalization as contrasted with weights, is nonadditive.

$$x_j = [s_1x_{1j}, s_2x_{2j}, ..., s_ix_{ij}, ... sx_{nj}] \quad \text{(II)}$$

with: s_i = the score of attribute i as determined by the normalization process $(i = 1,2,...,n;$ n the number of attributes)

x_j = the row vector of utility for alternative j

with: $j = 1,...,m;$ m the number of alternatives.

In this way a normalized matrix is obtained:

$[_Nx_{ij}]$ with: $i = 1,2,.....,n$ as the attributes

$j = 1,2,......,m$ as the alternatives

N indicates the normalization

$_Nx_{ij}$ = the normalized attribute i of alternative j.

The question could be asked, if even normalization can be dropped. For that reason multiplicative approaches are studied.

1.2 BILINEAR AND MULTILINEAR FORMS

Besides additive utilities, a utility function may also include a multiplication of the attributes. The two dimensional u(y,z) can then be expressed as a multilinear utility function[1]:

$$u(y,z) = k_yu_y(y) + k_zu_z(z) + k_{yz}u_y(y)u_z(z) \quad \text{(III)}$$

If $k_{yz} = 0$ we return to the additive form. For Keeney the additive form is rather a limiting case of the multiplicative utility function[2].

If $k_{yz} \neq 0$, then the utility function possesses a multiplicative part:

if $k_{yz} > 0$, then the mutual influence is positive,

if $k_{yz} < 0$, then the mutual influence has a negative effect on the utility function.

This representation mixes additive and multiplicative parts. It is not related to a multiplicative utility function nor to a product form, but to a bilinear representation of the form: $\Sigma_r\Sigma_s a_{rs}x_r y_s$. Indeed this representation is bilinear (and in general multilinear) and not purely multiplicative,

> "since two sets of variables are involved and each appears in a linear way.... and constant coefficients can be added to make the forms completely general"[3].

The danger exists that the multiplicative part becomes explosive. The multiplicative part of the equation would then dominate the additive part and finally would bias the results. It could happen if the factors are larger than 1, unless the weights for the multiplicative part are extremely low.

Considering these and the previous shortcomings, preference will be given to a method that is nonlinear, nonadditive, does not use weights and does not require normalization. Will a full-multiplicative form respond to all these conditions?

First, once again the alternatives have to pass the tests on lower bounds and upper limits and if some of them are not dominating. Indeed, filtering or domination may already solve the problem of optimization for different objectives. Consequently, we are only concerned with a ranking for the still incomparable alternatives at that moment.

1.3 THE FULL-MULTIPLICATIVE FORM

The following n-power form is called from now on *full-multiplicative form* in order to distinguish it from the previous mixed forms. It does not exclude that its factors may increase with a constant:

$$U_j = \prod_{i=1}^{n} (\alpha_i + \beta_i x_{ij}) \qquad \text{(IV)}$$

with: $j = 1,2,...,m$; m the number of alternatives
$i = 1,2,...,n$; n being the number of attributes and objectives
x_{ij} = response of alternative j on attribute i of objective i
α and β are measures of importance
U_j = overall utility of alternative j. U_j is a dimensionless indicator[4].

The outcome of this presentation is nonlinear, which presents an advantage, as the utility function of human behavior toward several objectives is certainly nonlinear. Several situations may occur, related to the α - terms and the β - coefficients.

1. All α-terms are zero but all β-coefficients are larger than zero

If three attributes and two alternatives are assumed:

$$n = 3 \qquad m = 2$$

a) in general:

$$\alpha_1 = \alpha_2 = \alpha_3 = 0$$
$$U_1 = \beta_1 \, \beta_2 \, \beta_3 \, x_{11} \, x_{21} \, x_{31}$$
$$U_2 = \beta_1 \, \beta_2 \, \beta_3 \, x_{12} \, x_{22} \, x_{32}$$

b) the β-coefficients have the form of weights:

$$0 \leq \beta_1 \leq 1 \quad 0 \leq \beta_2 \leq 1 \quad \beta_3 = 1 - \beta_1 - \beta_2$$

with the following outcome:

$$U_1 = \beta_1 \, \beta_2 (1 - \beta_1 - \beta_2) \, x_{11} \, x_{21} \, x_{31}$$
$$U_2 = \beta_1 \, \beta_2 (1 - \beta_1 - \beta_2) \, x_{12} \, x_{22} \, x_{32}$$

c) the β-coefficients have the form of ratios:

$\beta_1 = 1/\gamma_1 \qquad \beta_2 = 1/\gamma_2 \qquad \beta_3 = 1/\gamma_3$ (these ratios have the form of weights if they satisfy the same conditions as under b)

$$U_1 = (1/\gamma \;\; 1/\gamma_2 \;\; 1/\gamma_3) \, x_{11} \, x_{21} \, x_{31}$$
$$U_2 = (1/\gamma_1 \;\; 1/\gamma_2 \;\; 1/\gamma_3) \, x_{12} \, x_{22} \, x_{32}$$

Conclusion

In the three cases, the relation between the two overall utilities remains the same:

$$\frac{U_1}{U_2} = \frac{x_{11} \, x_{21} \, x_{31}}{x_{12} \, x_{22} \, x_{32}}$$

The following conclusions arise from the invariance between the overall utilities.

Rule I

In the full-multiplicative form the relation between the utilities does not change if more importance is given to an objective by multiplying it by a factor. Indeed, at that moment all alternatives are multiplied with that factor.

Consequence 1

In the full-multiplicative form the introduction of weights is meaningless. Indeed weights are here in fact multiplying coefficients.

Rule II

Rule II implies that in the full-multiplicative form an attribute of an objective can be divided by a constant without changing the relation between the overall utilities of the alternatives.

Consequence 2

In the full-multiplicative form an attribute of the size 10, 10^2, 10^3, 10^6, 10^9 etc. can be replaced by the unit size without changing the relationship between the utilities of the alternatives.

This consequence is extremely important for attributes expressed in monetary units. Instead of expressing an attribute in tens, hundreds, thousands, millions, billions for instance of dollars, the use of one digit in the integer part is sufficient.

2. All β-coefficients are equal to one but some or all α-terms are positive.

Alternatives with a smaller score in an attribute of an objective, but on which a positive α-term reacts, will be favored.

3. Some or all α-terms and all β-coefficients are larger than zero.

In this case, the importance of certain objectives is stressed by the β-coefficients, but the β-coefficients have no influence whatsoever. At the same time, alternatives with smaller scores in the attributes are favored by the α-terms, as in situation two.

1.4 WHAT ABOUT THE UNIT OF UTILITY U_J?

Until now, the units of the several attributes of the objectives (x_{ij} of formula IV) are taken into consideration as units of all kind. The same dimension per attribute is, however, present for all alternatives.

With the full-multiplicative form, the overall utilities (U_j), obtained by multiplication of different units of measurement, become dimensionless measures[5]. This situation would not bias the outcomes amidst the several alternatives as the last ones are represented by *dimensionally homogeneous equations*, being:

"formally independent of the choice of units"[6].

1.5 AN ILLUSTRATIVE APPLICATION

The example in Table 1.1 illustrates the theory. Three projects are competing in a firm having to maximize NPV (Net Present Value). At the same time, the government asks to maximize employment.

Table 1.1 Projects facing Multiple Objectives

	Increase of employment	NPV
Project 1	400 man-years	100 m. $
Project 2	300 man-years	150 m. $
Project 3	100 man-years	350 m. $

1) All α-terms are zero and all β-coefficients equal to one.
Applying Consequence 2 of Rule II the following outcome is obtained:

$$U_1 = 4 * 10 = 40$$
$$U_2 = 3 * 15 = 45$$
$$U_3 = 1 * 35 = 35$$

$$U_2 \text{ P } U_1 \text{ P } U_3$$

2) In addition to the previous case, assume that the stakeholders like to favor an attribute above the others. For instance, NPV is considered as valuable as two times the employment level:

$$U_1 = 4 * 2 * 10 = 80$$
$$U_2 = 3 * 2 * 15 = 90$$
$$U_3 = 1 * 2 * 35 = 70$$

$$U_2 \text{ P } U_1 \text{ P } U_3$$

The outcome remains unchanged but with all utilities two times higher than in the previous case (Rule I). Additionally one could conclude that in the full-multiplicative form the use of weights is senseless, as already mentioned in consequence 1.

3) All β-coefficients are equal to 1 and α_1 is equal to 100 person-years. It could be that the stakeholders favor a project by allowing lower labor productivity in that project.

$U_1 = 5 * 10 = 50$
$U_2 = 4 * 15 = 60$
$U_3 = 2 * 35 = 70$

$$U_3 \, P \, U_2 \, P \, U_1$$

As mentioned earlier, project U_3 with the smallest score in the attribute under consideration, namely employment, is favored by the introduction of an α-term for employment.

4) On the basis of Table 1.1, suppose that the stakeholders like to stress the NPV importance, for instance in the following way: all β-coefficients are equal to 1 and α_2 is put equal to 600 million dollars:

$U_1 = 4 * 70 = 280$
$U_2 = 3 * 75 = 225$
$U_3 = 1 * 95 = 95$

$$U_1 \, P \, U_2 \, P \, U_3$$

The outcome shows that a solution with the smallest score for NPV viz., U_1, may surpass the other solutions by introducing an α-term on NPV.

1.6 IS NORMALIZATION STILL NECESSARY ?

Contrary to weights, normalization is only meant for comparison and not for addition. Several mechanical devices for normalization are proposed.

1) One of the alternatives is taken as a reference[7]

If the first project of the illustrative example of Table 1.1 is taken as a reference, the following outcomes are obtained (the attributes of Project 1 (400,100) are used as denominators for the attributes of all projects):

$$U_1 = \frac{400}{400} * \frac{100}{100} = 1$$

$$U_2 = \frac{300}{400} * \frac{150}{100} = 1.125$$

$$U_3 = \frac{100}{400} * \frac{350}{100} = 0.875$$

$$U_2 \text{ P } U_1 \text{ P } U_3$$

This means the same outcome as before. Even the proportions between U_1, U_2 and U_3 did not change, as compared with the corresponding non-normalized case.

2) The attribute of each alternative is divided by the square root of the sum of the squares of all given values of that attribute[8]

We apply this normalization for the multiplicative form. The following results are obtained for the illustrative example:

$$U_1 = \frac{400}{\sqrt{16*10^4 + 9*10^4 + 10^4}} * \frac{100}{\sqrt{1*10^4 + 2.25*10^4 + 12.25*10^4}}$$

$$= \frac{400}{509.91} * \frac{100}{393.7} = \frac{4*10^4}{509.91*393.7}$$

$$U_2 = \frac{300}{509.91} * \frac{150}{393.7} = \frac{4.5*10^4}{509.91*393.7}$$

$$U_3 = \frac{100}{509.91} * \frac{350}{393.7} = \frac{3.5*10^4}{509.91*393.7}$$

$$U_2 \text{ P } U_1 \text{ P } U_3$$

The proportions did not change in comparison with the original outcomes with α-variables equal to zero and β-coefficients equal to one.

3) The division of the attributes per alternative by their maximum value[9]

We apply this normalization for the multiplicative form:

$$U_j = \frac{x_{1j}}{x_1^+} * \frac{x_{2j}}{x_2^+} * \ldots * \frac{x_{ij}}{x_i^+} * \ldots * \frac{x_{nj}}{x_n^+}$$

with x_i^+ defined as the maximum x_{ij}

Application on the illustrative example gives:

$$U_1 = \frac{400}{400} * \frac{100}{350} = \frac{40{,}000}{400*350}$$

$$U_2 = \frac{300}{400} * \frac{150}{350} = \frac{45{,}000}{400*350}$$

$$U_3 = \frac{100}{400} * \frac{350}{350} = \frac{35{,}000}{400*350}$$

i.e., $U_2 \, P \, U_1 \, P \, U_3$

with the same proportions as in the non-normalized case with all α-variables equal to zero and all β-coefficients equal to one.

4) Normalization through division by the column total[10]

Still another kind of normalization is possible by dividing each attribute by the sum of all given values of that attribute, i.e., for the full-multiplicative form and for the given example:

$$U_1 = \frac{400}{800} * \frac{100}{600}$$

$$U_2 = \frac{300}{800} * \frac{150}{600}$$

$$U_3 = \frac{100}{800} * \frac{350}{600}$$

i.e., $U_2 \, P \, U_1 \, P \, U_3$

shows the same proportions as in the non-normalized example.

Consequence 3

Normalization of the full-multiplicative form is meaningless.
This was already concluded for weights (see Consequence 1).

1.7 OBJECTIVES MOVING IN A DIFFERENT DIRECTION

Until now, all objectives were implicitly assumed to move in the same direction. No problem arises if all objectives are maximized, which is also the case if all objectives are minimized. A difficulty emerges, if some objectives have to be maximized and some others to be minimized.

If in an example of a firm a break-even point is reached e.g., after two years in a first alternative, but after three years in a second one and after three years and six months in a third one, then the first alternative is the best for this objective.

How is it possible to combine this minimization problem with the maximization of the other objectives? Therefore, the objectives to be minimized are put as denominators in the formula:

$$U_j' = \frac{A_j}{B_j} \qquad \text{(V)}$$

with: $A_j = \prod_{g=1}^{i} (\alpha_g + x_{gj})$

$j = 1,2,...,m;$ *m* the number of alternatives

i = the number of objectives to be maximized

with: $B_j = \prod_{k=i+1}^{n} (x_{kj} - \alpha_k)$

$n\text{-}i$ = the number of objectives to be minimized.

with: U_j' = the utility of alternative j with attributes to be maximized and attributes to be minimized.

The α-terms are provided with a negative sign in the minimization process.

Aside from the α-terms, the introduction of the β-parameters would not change the result of the formula. Indeed the multiplications: $(\beta_1.\beta_2.....\beta_i)$ and $(\beta_{i+1}......\beta_n)$ would be present in all the alternatives in the same way as was indicated above.

This way of operating is completely logical. Multiplication and division are fundamental arithmetic operations, strongly linked to each other. Consequently division and quotient of quantities take their place in the model.

1.8 AN ILLUSTRATIVE EXAMPLE WITH OBJECTIVES MOVING IN A DIFFERENT DIRECTION

Three projects are competing in a firm having to maximize NPV and to minimize the *Break-even Period*. At the same time, the government asks to maximize employment (Table 1.2).

Table 1.2 Examples of Multiple Objectives Moving in a Different Direction

Alternatives	Increase of employment	NPV	Break-even period
Project 1	400 man-years	100 m. $	2 years
Project 2	300 man-years	150 m. $	3 years
Project 3	100 man-years	350 m. $	3.5 years

1) All α-terms are zero and all β-coefficients equal to 1:

$U_1 = (4 * 10): 2 = 20$
$U_2 = (3 * 15): 3 = 15$
$U_3 = (1 * 35):3.5 = 10$

$$U_1 \; P \; U_2 \; P \; U_3$$

2) In addition to the previous case, an attribute is favored above the others. For instance, NPV is assumed as valuable as two times the employment level:

$U_1 = (4 * 2 * 10): 2 = 40$
$U_2 = (3 * 2 * 15): 3 = 30$
$U_3 = (1 * 2 * 35):3.5 = 20$

$$U_1 \; P \; U_2 \; P \; U_3$$

The outcome remains unchanged as compared with the previous case, concerning not only the ranking, but concerning also the proportions between the utilities.

Additionally in the full-multiplicative form, with objectives moving in a different direction, once again the use of weights makes no sense.

3) All β -coefficients are equal to 1, but α_1 = 100 man-years:

$U_1 = (5 * 10): 2 = 25$
$U_2 = (4 * 15): 3 = 20$
$U_3 = (2 * 35):3.5 = 20$

$$U_1 \; P \; U_2 \; I \; U_3$$

As mentioned earlier, project U_3 with the smallest score for the attribute under consideration, namely man-years, is favored.

1.9 MULTIPLICATIVE FORM ON THE BASIS OF GEOMETRIC MEANS

It is quite normal that the arithmetic average has to be used for the additive form. Consequently, the geometric mean is more suitable for the multiplicative form. Concerning ratings Roberts has demonstrated that the geometric mean is superior to the arithmetic one[11].

The ranking, however, will not change if the geometric mean is applied on formula (IV). This is also the case with formula (V) as algebraically:

$$\frac{\sqrt[n]{A}}{\sqrt[n]{B}} = \sqrt[n]{\frac{A}{B}} \qquad \text{(VI)}$$

This means that formulae (IV) and (V) can be maintained with the mentioned rules and consequences.

1.10 THE DANGER OF A REVERSE ORDER IN THE MULTIPLICATIVE FORM

With multilateral comparisons of several alternatives, the possible contradictory outcome, typical for two by two comparisons,(cf. Paradox of Condorcet[12]) is out of the question.

It is not so sure concerning a possible Reverse Order. Indeed a reverse order (cf.Arrow's General Possibility Theorem[13]) is possible, if a new alternative is introduced.

The new alternative will not influence the positions of the existing alternatives to one another, as their responses are not affected. Therefore, we return to the example: objectives moving in a different direction (see Table 1.2).

A strong new project (project 4) is introduced later, with as scores for the attributes:

400 man-years of employment
350 m. $ of NPV
5 years break-even.

1) All α-terms are zero and the β-coefficients equal to 1:

$U_1 = 20$ $U_2 = 15$ $U_3 = 10$

$U_4 = (4 * 35): 5 = 140 : 5 = 28$

$$U_4 \text{ P } U_1 \text{ P } U_2 \text{ P } U_3$$

2) All β-coefficients are equal to 1 and α_1 = 100 man-years (as in point 3 after Table 1.2):

$U_1 = 25$ $U_2 = 20$ $U_3 = 20$

$U_4 = (5 * 35): 5 = 175 : 5 = 35$

$$U_4 \text{ P } U_1 \text{ P } U_2 \text{ I } U_3$$

In the two examples, the positions of the old projects are maintained though dominated by the new project.

What does happen if a new project is weaker than some of the existing projects? A new project (project 4) is introduced later, this time with as scores for the attributes:

300 man-years of employment
7 years break even

1) All α-terms are zero and the β-coefficients equal to 1:

$U_1 = 20$ $U_2 = 15$ $U_3 = 10$

$U_4 = (3 * 35): 7 = 105 : 7 = 15$

$$U_1 \text{ P } U_2 \text{ I } U_4 \text{ P } U_3$$

2) All β-coefficients are equal to one and α_1 = 100 man-years:

$U_1 = 25$ $U_2 = 20$ $U_3 = 20$

$U_4 = (4 * 35): 7 = 140 : 7 = 20$

$$U_1 \text{ P } U_2 \text{ I } U_3 \text{ I } U_4$$

i.e., the positions of the old projects are maintained; only the new project is dominated by project 1 and indifferent to projects 2 and 3.

1.11 ATTRIBUTES OF OBJECTIVES RAISED TO POWERS

Allocating an exponent to an attribute of an objective signifies stressing the importance of this attribute. Formula V is adjusted for that purpose:

$$U_j'' = \frac{A_j''}{B_j''} \qquad \text{(VII)}$$

$$\text{with: } A''_j = \prod_{g=1}^{i} (\alpha_g + x_{gj})^{\varepsilon g}$$

$j = 1,2,...,m;$ m the number of alternatives

i = the number of attributes (objectives) to be maximized

α_g = α-parameter for attribute (objective) g

$(\alpha_g + x_{gj})^{\varepsilon g}$ = response of alternative j on attribute (objective) g supplied with exponent εg, which is linked to attribute g.

The additive parameters α make a logarithmic transformation impossible[14];

$$\text{with: } B''_j = \prod_{k=i+1}^{n} (x_{kj} - \alpha_k)^{\varepsilon k}$$

$n - i$ = the number of attributes (objectives) to be minimized.

α_k = the α-term for attribute k, has to be supplemented with a negative sign.

$(x_{kj} - \alpha_k)^{\varepsilon k}$ = response of alternative j on attribute (objective) k supplied with exponent εk, which is linked to attribute k

with: U''_j = utility of alternative j with some attributes raised to powers.

U''_j is a dimensionless indicator.

1) The Use of One Digit in the Integer Part

Is consequence 2 of Rule II also of application here? Miller and Starr[15] bring the answer. Table 1.3 demonstrates ratios that do not change if single-dollar units are transformed to million-dollar units even when attributes are raised to powers.

The costs, which are expressed in dollars, such as investment and labor costs, have to be minimized. Therefore, the other objectives are minimized too: 0 for the best and 10 for the poorest (see Table 1.3).

Table 1.3 Example of Multiple Objectives with Exponents

Attributes	Units	Scores	Strategy S1	Strategy S2
Investment	Dollars	3	1,000,000	1,500,000
Site	Preference*	4	6	10
Labor cost	Dollar/hour	5	1.7	1.5
Relations	Cooperation*	1	7	4
Raw-material	Quality*	3	2	8
Transport	Facility*	2	9	6

* 10 poorest; 0 best.

"An Incorrect Method That Is Frequently Used"[16]

An additive method with the scores as weights gives different results by changing the scale:

a) the dollar-attributes are expressed in unit dollars: S2/S1 = 1.50

b) in million dollars units instead of single-dollar units.
The sum of the products of the scores (here weights) with the strategies changes: S2/S1 = 1.38

The change in scale introduces dimensional distortion in the ratio.

A Dimensionally Correct Method:

a) the dollar-attributes are expressed in unit dollars:

$$S1 = (1{,}000{,}000)^3(6)^4(1.70)^5(7)^1(2)^3(9)^2 = 8.34 * 10^{25}$$
$$S2 = (1{,}500{,}000)^3(10)^4(1.50)^5(4)^1(8)^3(6)^2 = 1.89 * 10^{28}$$

$$\frac{S2}{S1} = \frac{1.89 * 10^{28}}{8.34 * 10^{25}} = 226$$

b) in million dollars units:

$$S1 = (1)^3(6)^4(1.70)^5(7)^1(2)^3(9)^2 = 8.34 * 10^{7}$$
$$S2 = (1.5)^3(10)^4(1.50)^5(4)^1(8)^3(6)^2 = 1.89 * 10^{10}$$

$$\frac{S2}{S1} = \frac{1.89 * 10^{10}}{8.34 * 10^{7}} = 226$$

S1 is preferred to S2 (a minimization process).

The ratios are indeed invariant to the transformation of scale. The indicator 226 is dimensionless and of an ordinal order. It does certainly not mean that S1 is 226-times better than S2.

2) Objectives moving in a Different Direction

In Miller and Starr's method costs are minimized. The introduction of another scale (0 - 10) for the attributes to be maximized (site, relations, raw-material, transport) is not very elegant.

To give an opposite scale to the attributes to be minimized is another possibility. For instance in Table 1.2 the break-even period could read:

Originally	Opposite scale
2 years	3.5
3 years	3
3.5 years	2

The method with formulas (V) and (VII) is preferable above these other methods.

3) Contrary to the β-coefficients the introduction of Exponents will influence the Ranking of the Alternatives

Returning once again to the example of Table 1.2 point 1, where the following ranking was obtained:

$U_1 = 20; U_2 = 15; U_3 = 10$:

$$U_1 \; P \; U_2 \; P \; U_3$$

If an exponent 2 is given to N.P.V the ranking changes:

$$U_1 = (4 * 10^2): 2 = 400 : 2 = 200$$
$$U_2 = (3 * 15^2): 3 = 675 : 3 = 225$$
$$U_3 = (1 * 35^2):3.5 = 1225 : 3.5 = 350$$

$$U_3 \; P \; U_2 \; P \; U_1$$

4) How to stress the Importance of an Objective?

In order to stress the importance of an objective, a range of possibilities is offered, either by increasing its exponent or eventually through its α-term.

For the application of the economic law of decreasing marginal utility, an interval range is introduced, whereby the exponents are decreased and if necessary even expressed as a fraction. It is also possible to decrease the α-terms or even to provide them with a minus sign.

5) Awarding Exponents or eventually α-terms to Attributes may introduce Subjectivity, which has to be avoided as much as possible

Either in a heuristic way or by a Delphi exercise, subjectivity can be avoided. With the heuristic approach, the argumentation has to satisfy all the stakeholders. This involvement of all the stakeholders is later illustrated in an application for an optimal economic policy for Belgium.

Delphi on the other hand tries to reach all possible stakeholders represented through especially knowledgeable people. Additionally the method is based on anonymity, avoiding bandwagon effects typical for committees and meetings. Numerous feedbacks try to reach convergence as much as possible. In this way the method strives after objectivity (for more details on Delphi, return to Part II, 1.3 of this book). Nominal Group Technique is comparable to Delphi but less costly, faster and simpler. Nominal Group Technique can be used independently or as a first approximation for Delphi (for more details, return to Part II, 1.4 and Chapter 1, Annexes 1, 2, and 3 of this book).

1.12 SOME RESTRICTIONS OF THE MULTIPLICATIVE REPRESENTATION

The multiplicative representation is only applicable if the alternatives are known in detail beforehand (the *Discrete Case*) and not for alternatives that are automatically generated during the process (the *Continuous Case*). It has to be stressed that at the start the multiplicative representation works with *Cardinal Numbers*. Nevertheless, the finally obtained dimensionless numbers are compared in an *Ordinal Scale*. The stakeholders are mainly interested in a problem of choice, in particular social choice. Social choice shows interest in ranking. Consequently, an ordinal outcome does not present a problem[17].

The application of the full-multiplicative form is restricted to alternatives striving after the same objectives with their connected attributes. A problem would arise if one or more attributes were missing in an alternative. If the presence of these attributes forms a necessary condition as hard constraints, the alternative is rejected right away during a preceding filtering stage. The obligation to increase employment could be a hard constraint. If at that moment an alternative applies automation without increase in employment, this alternative is immediately dropped. If the increase in employment is only a soft constraint, the alternative in question is maintained but with a zero result in employment. A zero factor in a multiplication, however, means a zero outcome in the multiplicative form. This nonsense situation is corrected by giving to the missing attribute an extremely low symbolic value, as near as possible to zero.

NOTES PART FIVE CHAPTER 1

[1] R.L.Keeney, H.Raiffa, Decisions with Multiple Objectives. Preferences and Value Tradeoffs, Cambridge, University Press, USA, 1993, 234.

[2] R.L.Keeney, *A decision analysis with multiple objectives: the Mexico City Airport*, the Bell Journal of Economics and Management Science, vol.4.N° 1,1973, 110.

[3] R.G.D.Allen, *Mathematical Economics*,Macmillan,1957, 473.

[4] The formula was inspired by Schärlig. He considers, however, the utility of each response of an alternative on an attribute, which means that he has to normalize all responses. On the contrary, we assume that usefulness is included in the responses themselves. Indeed in the public sector it is accepted that the public authority can judge the utility of the attributes for all the stakeholders by indicators, norms etc. This happens in traditional government governance systems. In addition, the private sector has to operate in the same way (see therefore Part III, Chapter 1 of this book). A.Schärlig, *Décider sur plusieurs critères*, Presses Polytechniques Romandes, Paris, 1985, 87.

[5] Bridgman discussed dimensionless measures already in 1922. P.W.Bridgman,. *Dimensional Analysis*, Yale University Press, New Haven 1922 (new ed.1963).

[6] F.J.De Jong, *Dimensional Analysis for Economists*, North-Holland, Amsterdam,1967, 28.

[7] A.Schärlig, *Décider sur plusieurs critères*, Presses Polytechniques Romandes, Paris,1985, 76.

[8] A.Van Delft, P.Nijkamp, *Multi-criteria Analysis and Regional Decision-making*, M. Nijhoff, Leiden, Nl.1977, 37.

[9] A.Van Delft, P.Nijkamp, *op. cit.*, 38.

[10] H.Voogd, *Multicriteria Evaluation for Urban and Regional Planning*, Pion ltd., London,1983, 86-87.

[11] F.S.Roberts, Limitations on Conclusions using Scales of Measurement, in: S.M.Pollock et al.(eds), *Handbooks in OR and MS*, Vol.6,chapter 18, 1994,Elsevier, New York, 632.

[12] Marquis de Condorcet, *Essai sur l'application de l'analyse à la probabilité des décisions rendues à la pluralité des voix*, l'Imprimerie Royale, Paris, 1785, LXI.
W.K.Brauers, *Different concepts in multiobjective optimization*, Working Paper 99/12, RUCA, University of Antwerp, Faculty of Applied Economics, Antwerp,1999.
See also: Part III, 2.5 of this book.

[13] K.J.Arrow, *Social Choice and Individual Values*, Yale University Press, New Haven (2nd edit.) 1963, 13, 15, 97.
G.Bordes, and N.Tideman,. Independence of irrelevant alternatives in the theory of voting, Theory and Decision, Kluwer, Dordrecht(Nl.), 1991, 2, 183.
See also: Part III, 2.5 of this book

[14] Indeed such a logarithmic transformation is biased cf.P.Kennedy, *A Guide to Econometrics*, Fourth Edition, Blackwell, Oxford, 1998, 97-98; 115.
A.S.Goldberger, *The Interpretation and Estimation of Cobb-Douglas Functions*, Econometrica, Journal of the Econometric Society, Amsterdam, 1968, 36, 464-472.

[15] D.W.Miller, M.K.Starr, *Executive Decisions and Operations Research*, 2nd edition, Prentice-Hall Inc. Englewood Cliffs(N.J.)1969, 237-239.

[16] D.W.Miller, M.K.Starr, *op. cit.*, 238.

[17] See for more details: K.J.Arrow, *op. cit.*
H.S.Hargreaves; M.Hollis, B.Lyons, R.Sugden, A.Weale, *The Theory of Choice*, A Critical Guide, Blackwell, Oxford, 1992.

CHAPTER 2

THE MULTIPLICATIVE REPRESENTATION IN SCENARIOS FOR AN OPTIMAL ECONOMIC POLICY IN BELGIUM

The determination of an optimal economic policy for Belgium constitutes an application of the multiplicative representation. Different scenarios will represent hereby the alternatives[1].

What is the importance of scenario writing in this context? Scenario writing provides the opportunity to compare several alternative economic policies. Moreover, scenario writing tries to cope with the uncertainty of the future (for more details about scenario writing return to Part II, Chapter 3 of this book).

Scenarios are strictly advisory and not deterministic. Completeness represents another characteristic of scenario writing. Only a very limited number of scenarios should be considered, unless the mutual consistency of the scenarios is computer checked. The presentation of eleven scenarios for an optimal Belgian economic policy maintains this philosophy, thanks to a computer assisted multiplicative representation. A weak point could be that scenarios are characterized by a lack of interaction between the events. We will especially take care that also interaction occurs.

An insight in the political structure of Belgium is necessary in order to have a better understanding of this application.

2.1 POLITICAL AND ECONOMIC STRUCTURE OF BELGIUM

Belgium is a federal state, composed of three regions: Flanders, Wallonia and the capital Brussels. The optimal economic policy for Belgium forms the top of a top down approach with as local optima the economic growth of the separate Belgian regions. Descending lower will increase specificity. This specificity finally boils down to measurability, ending up with several measurable subobjectives viz., maximization of private income per head in each of the regions, minimization of transfer payments from the richest region (Flanders) and minimization of the national debt. Indeed the European Union considers the Belgian national debt as being too high. This national debt could be carried by the regions together or could be split up over the regions.

For the Belgian application of scenarios, is there a lack of interaction? How have the subobjectives been selected and have others been ignored? Does the danger exist that maximization of private income maximizes consumption instead of savings and investment? The propensity of the Belgians to save is already one of the highest in the world and the financing of the national debt occurs mainly by the Belgians themselves. Minimization of the national debt will liberate more money for private investment.

Another possibility to consider could be maximization of employment. Already nearly full employment exists in Flanders. Lower subsidies for Wallonia and Brussels, by decreasing transfer payments, will push them to be more active. This will diminish the hidden unemployment in the public sector that exists in these regions.

The Belgian balance of payments has shown a surplus for many years. Maximization of this surplus would show no financial solidarity with other nations. Indeed ipso facto deficits in their balances of payments would be created, for instance for developing countries.

Minimization of inflation is within the competence of the European Central Bank. The ECB tries to keep the inflation rate under the 2% level.

This exposition shows sufficiently, that, within the Belgian competence, interactions are included in the eleven Belgian scenarios.

2.2 ELEVEN SCENARIOS FOR AN OPTIMAL ECONOMIC GROWTH IN BELGIUM

Scenario 0

This scenario represents the existing situation. The richest region, Flanders, reaches the limit concerning subsidies towards the other regions of Belgium, Wallonia and Brussels[2]. An *Income Paradox* is the unexpected result, namely that through the subsidies the private income of the average Flemish becomes lower than that of the average Walloon or the average inhabitant of Brussels. Consequently, it cannot be maintained and is therefore called scenario 0.

Scenario I

Scenario I stands for the scenario of the average Belgian: the subsidies are limited in such a way that the average income of the average Belgian is the same everywhere, whereas the public debt is really kept national.

Scenario II

Scenario II assumes a zero growth in Flanders. From general international experience, a limit for transfer payments (subsidies) is accepted of 3.3% of *Gross Regional Product (GRP)* under zero growth and no inflation. In this scenario, the limit for transfer payments is included together with zero growth in Flanders.

Scenario III

In scenario III the national public debt is split over the regions. This split represents a disadvantage for the Walloon and Brussels regions.

Scenario IV

Scenario IV moves further than scenario II by also splitting the national debt over the regions.

Scenario V

Scenario V includes the scenario of the average Belgian plus the split of the national debt over the regions.

Scenario VI

Scenario VI includes an exceptional Flemish goodwill to the other regions, if the national debt is split. In addition, a preliminary agreement between the regions is supposed to exist, involving rather a confederate structure instead of the actual Belgian federal structure.

Scenario VII

Financial solidarity between the regions is maintained through keeping the public debt national and not through the transfer payments, which are abolished.

Scenario VIII

Scenario VIII covers the separation of the Belgian regions with splitting of the national debt, whereas subsidizing occurs through European funds.

Scenario IX

Scenario IX includes the economic revival of Wallonia and Brussels by bringing these regions on the level of the Flemish growth.

The average incomes of Wallonia and Brussels are simply assumed as high as in Flanders. It is true that it will perhaps never be the case, but that is not the point. The point is that an ideal optimum would be reached for the regions and for Belgium if, without subsidies, i.e., transfer payments, Wallonia and Brussels would be economically so productive as Flanders.

The original situation is comparable to a country whose industry is inefficient by being protected with very high tariffs. Liberation of tariffs and abolition of subsidies would be healthy for that country in the long run as was proven so many times.

These remarks lead to an interesting conclusion for a general regional policy of the nations. Economic growth between the regions can be subject to three possibilities. First, concerning average incomes, the richest region is brought down to the level of the poorer regions. This may happen in scenario 0. Second, the richest region is brought down to an average national level and the poorer regions are lifted up to that average. Scenario I translates this possibility. Finally, in Scenario IX, the poorer regions are raised to the level of the richest region.

Scenario X

Scenario X includes the previous scenario plus the split of the national debt.

The Data

Table 2.1 gives the data for the several scenarios on the basis of the regional statistics for the year 1996 (in 1996 prices)[3]. Data that are more recent are not available. No relative changes between the regions can be assumed for the recent years. Incomes and subsidies are expressed in the previous Belgian franc (BEF), with the Belgian franc fluctuating around 40 BEF for one American dollar. The public debt represents a percentage of the Gross Regional Product (GRP) of the regions.

2.3 THE RANKING AFTER THE APPLICATION OF THE MULTIPLICATIVE REPRESENTATION

A problem would arise if one or more attributes were missing in an alternative. It is the case in one of the columns of scenarios VII to X (see table 2.1).

If the presence of these attributes forms an essential condition, they have to be rejected right away. Otherwise, they are maintained, but with a multiplication factor of zero, they would be eliminated anyway. Therefore, a symbolic value of one Belgian franc (representing approximately 2.7 dollar cents) was given to the missing attributes or 1/19,139th (0.0000522th) part of the smallest corresponding attribute (using the BEF-cent as a basis the proportion would be 0.000000522).

Of course, the alternative with this missing link will start with a handicap in the maximization process. On the contrary, in the minimization process it would mean an advantage. As said, scenarios VII to X enjoy that advantage (see Table 2.1).

Table 2.1 Scenarios for an Optimal Economic Policy in Belgium

Scenarios	Private Income per head, Wallonia	Private Income per head, Brussels	Private Income per head, Flanders	Subsidies/ head from Flanders	Public Debt of Flanders	Public Debt of Wallonia	Public Debt of Brussels
	MAX. (in BEF)	MAX. (in BEF)	MAX. .(in BEF)	MIN. (in BEF)	MIN. (% GRP)	MIN. (% GRP)	MIN. % GRP)
Scenario O	692,883	698,809	676,743	41,771	122.2	122.2	122.2
Scenario I	684,076	684,076	684,076	34,438	122.2	122.2	122.2
Scenario II	657,935	680,457	699,375	19,139	122.2	122.2	122.2
Scenario III	674,574	699,861	686,881	31,633	116.4	134.6	120.6
Scenario IV	657,935	680,457	699,375	19,139	116.4	134.6	120.6
Scenario V	684,076	684,076	684,076	34,438	116.4	134.6	120.6
Scenario VI	692,883	698,809	676,743	41,771	116.4	134.6	120.6
Scenario VII	628,380	664,936	718,514	--	122.2	122.2	122.2
ScenarioVIII	628,380	664,936	718,514	--	116.4	134.6	120.6
Scenario IX	718,514	718,514	718,514	--	122.2	122.2	122.2
Scenario X	718,514	718,514	718,514	--	116.4	134.6	120.6

Based on the data of Table 2.1, formula V of the multiplicative representation is applied. The optimal economic policy is deducted from a ranking in decreasing order of the indicators from the different scenarios. As a characteristic of the multiplicative model, the indicators may show very high differences. These differences are certainly not acceptable as cardinal numbers differences. The scale of the indicators is although better than an *Ordinal Scale*. Indeed, the scale of the indicators is very revealing as a rough indication of relative distances. After this interpretation, the scenarios ranking from 1 until 4 are highly superior to the other ones (see Table 2.2).

Table 2.2 Ranking of the Scenarios with the Multiplicative Form

1	Scen.IX	Economic revival of Wallonia and Brussels	203,267
2	Scen.X	Economic revival of Wallonia and Brussels plus the split of the national debt	196,306
3	Scen.VII	Financial solidarity only through the national debt kept really national	164,515
4	Scen.VIII	Secession	158,881
5	Scen.II	Zero growth in Flanders	90
6	Scen.IV	Zero growth in Flanders plus the split of the national debt	87
7	Scen.III	The split of the national debt	54
8	Scen.I	Scenario of the average Belgian	51
9	Scen.V	Scenario of the average Belgian plus the split of the national debt	49
10	Scen.O	Status quo 1996	43
11	Scen.VI	Flemish goodwill in the split of the national debt	42

An optimal economic policy for Belgium depends, in the first place, on an economic revival of Wallonia and Brussels, including the end of transfer payments coming from Flanders. Could not the end of subsidizing with self-sufficiency, in social security in particular, stimulate a dynamical Wallonia and a dynamical Brussels?

Moreover, financial solidarity through transfer payments does not seem very effective for Belgian public management, being rather a source of paralyzing immobility. On the contrary, financial solidarity, through keeping the national public debt really national, is effective. Moreover splitting of the national debt would mean a heavy burden for Wallonia. A temporary aid from Flanders for this purpose would not be very helpful.

Finally, in the light of European Unity and European solidarity, and only considering economic reasons, the splitting up of Belgium itself would not be such a great disaster. In addition, either the scenario of the average Belgian or the scenario of Flemish goodwill in a more confederate structure, are not very welcome.

2.4 IS IT POSSIBLE TO STRESS THE IMPORTANCE OF SOME OBJECTIVES?

In this section, scores are tested in order to show that more importance is given to one or more objectives. It concerns the α-variables, the β-coefficients and the attributes raised to powers.

The objective of the minimization of the public debt on the national level, or distributed over the regions, will be the subject of this test. The Maastricht-norm of May 1997 imposed the reduction of the Belgian public debt, at that moment at 122.2% of the *Gross Domestic Product (GDP)*, to 60%. If the public debt of 122.2% is regionalized, 116.4% of its Gross Regional Product would affect Flanders, 120.6% Brussels and 134.6% Wallonia. Table 2.1 (columns 6 to 8) records these figures

1) The relevance of the α-variables in a new simulation

All the figures of the public debt that were given above are diminished by 60, i.e., the 60% that is allowed at the European level. In this way, the α-variables are provided with a negative sign, which is normal for minimization. Despite this change, the ranking of the variables remained unchanged.

2) The relevance of the β-coefficients

If reduction of the public debt to the 60%-norm would take five years, a β-coefficient of 5 is allotted to the public debt. As was already demonstrated in the theoretical part of this report, the ranking of the scenarios cannot and does not change.

3) The relevance of attributes raised to powers

This relevance is more fundamental. Consequently, it is discussed more thoroughly.

2.5 RELEVANCE OF ATTRIBUTES RAISED TO POWERS FOR AN OPTIMAL BELGIAN ECONOMIC POLICY

There exists no fundamental reason for raising to powers private incomes and transfer payments. Indeed the different scenarios already show multiple variations of these objectives.

For the public debt, it is not the same Indeed the burden of the public debt remains a nightmare for Belgium for many years to come, a handicap unknown for the other competitive members of the European Union with exception of Italy. Initially, the Belgian Planning Office estimated, that the necessary diminution of the public debt to 60% of GDP would be halfway over five years[4]. Consequently, by linear extrapolation, the target of 60% would be reached over ten years.

For all these reasons all stakeholders will agree that exponents of 5 and 10 are acceptable for the Belgian public debt. Therefore, instead of a Delphi exercise, which was suggested earlier, a heuristic path is followed (Table 2.3).

Table 2.3 Ranking of Scenarios influenced by the Public Debt over Five Years

Rank public debt over 5 years	Indicators	Scenarios	Title Scenarios	Rank public debt over 1 year
1	1,833.21	Scen.IX	Economic revival of Wallonia and Brussels	1
2	1,540.11	Scen.X	Economic revival of Wallonia and Brussels +the split of the national debt	2
3	1,483.72	Scen.VII	Financial solidarity only through the national debt kept really national	3
4	1,246.50	Scen.VIII	Secession	4
5	8.09	Scen.II	Zero growth in Flanders	5
6	6.79	Scen.IV	Zero growth in Flanders plus split of the national debt	6
7	4.59	Scen.I	Scenario of the average Belgian	8
8	4.26	Scen.III	Split of the national debt	7
9	3.88	Scen.O	Status quo 1996	10
10	3.86	Scen.V	Scenario of the average Belgian plus split of the national debt	9
11	3.26	Scen.VI	Flemish goodwill in the split of the national debt	11

What conclusions can be drawn from the comparison between the ranking with or without exponents?

The change in the dimensionless indicators does not permit an absolute comparison between the two cases. At the most, their size may give an indication, namely that the general decrease of the indicators could show a decrease in an optimal economic horizon for Belgium. This diminution is understandable, as the ranking without exponents did not take into account the heavy burden of the public debt for the following years. On the other side, the ranking does not essentially change by the introduction of exponents. Only at the bottom of the ranking some changes occur, especially a further retrogression of the regional split of the national debt.

What is the result, if the exponent of the public debt increases from 5 to 10? Indeed, it would need ten years for Belgium to reach the 60% norm. Therefore, in Table 2.4 an exponent of 10 is given to the national debt.

This Table 2.4 shows once again a general diminution of the indicators in comparison with the previous table. This indicates an economic punishment of Belgium by the European Union over a period of ten years. This situation is clearly underestimated in Belgium.

The changes in the ranking are not essential. This is also understandable as the allocation of exponents is the same for all scenarios. If the economic revival of Wallonia and Brussels remains in the first place, the second rank about the liquidation of the subsidies is heavily connected to it. Keeping the public debt nationally is the general tendency, whereas the split of the national debt is increasingly disapproved.

Table 2.4 Ranking of Scenarios with Public Debt amortized until the European Norm over Ten Years

Rank public debt 10 year.	Indicators	Scenario	Scenario Description	Rank public debt 5 year	Rank public debt 1 year
1	90.6033	Scen.IX	Economic revival of Wallonia and Brussels	1	1
2	73.3302	Scen.VII	Financial solidarity only through the national debt kept really national	3	3
3	63.9480	Scen.X	Economic revival of Wallonia and Brussels + split of the national debt	2	2
4	51.7566	Scen..VIII	Secession	4	4
5	0.3996	Scen.II	Zero growth in Flanders	5	5
6	0.2824	Scen.IV	Zero growth in Flanders plus split of the national debt	6	6
7	0.2271	Scen.I	Scenario of the average Belgian	7	8
8	0.1916	Scen.O	Status quo 1996	9	10
9	0.1768	Scen.III	Split of the national debt	8	7
10	0.1603	Scen.V	Scenario of the average Belgian plus the split of the national debt	10	9
11	0.1352	Scen.VI	Split of the national debt	11	11

A result for an optimal economic growth for Belgium and its regions is found after several simulation exercises, using scenarios. The increase in economic activity in the Belgian regions lagging behind, if based on self-help and not on transfer payments from the richer region, ranks first.

Objections can be made about the fact that in this optimal scenario the less wealthy regions will approach the income of the richest region. It is quite clear that at that moment the Belgian economic growth will reach an optimum, all other variables remaining constant. The idea behind the contents of this scenario is that the less wealthy regions would consider the income of the richer region as an *Ideal Point* to be reached as near as possible. Even if their income would rise halfway, they would be better off than in all the other remaining scenarios. Indeed , at that moment, the private income per head in Flanders being 718,514BEF would become 673,448BEF in Wallonia and 691,725BEF in Brussels. With the national debt amortized over ten years to the European norm, this scenario would get an indicator of 81.759, ranking second immediately after scenario IX. For the Belgian situation anyhow, it is far more preferable than the scenario where the richest region is brought down to an average income and the less wealthy regions are lifted up to that average. Scenario I illustrates this situation, but lagging far behind with an indicator of 0.2271.

If all this will happen without further subsidies (transfer payments), financial solidarity remains by keeping the national debt national, which means a higher contribution from the richer Flemish region. If these self-evident propositions are not taken for granted, transfer payments from the richer region, Flanders, to the poorer regions have to be diminished at least. Indeed the Flemish region cannot afford anymore to contribute above a certain ceiling. What's more, the national debt will be split anyway and even secession is threatening in the light of a stronger Europe and weaker national participating states.

2.6 CONCLUSIONS

The multiobjective multiplicative form was used in order to rank effects of alternative scenarios for Belgian economic development. A conclusive result is found after a ranking of eleven possible scenarios. The ninth scenario dominates all the others. This is still the case after changing the importance of the objectives through raising them to powers in a heuristic approach. Even then, three other scenarios remain relatively important. The other remaining scenarios are considerably lagging behind. Consequently it is clear that also a second best, a third best and even a fourth best scenario can be selected. In this way, the stakeholders could eventually make a choice between the next best scenarios. Indeed, it has to be stressed that scenario writing, like every other simulation exercise, is only indicative and not deterministic. After all scenario writing represents an instrument of creative thinking.

NOTES PART FIVE CHAPTER 2

[1] The report on this subject written in Dutch was one of the papers, which was presented for information to the new Belgian government in 1999, namely:
W.K.Brauers, *Een optimaal economisch beleid voor België en zijn gewesten, volgens de multipele effectiviteitsanalyse*, SDD Documentatieblad, Department of Finance Belgium, LIX, 5,1999,3-24.
The other paper was: W.K.Brauers, *Het Bruto Regionale Product van Vlaanderen, Wallonië en Brussel*, Working Paper 99/2, University of Antwerp (RUCA), Department of Applied Economics, 1999.

[2] W.K.Brauers, *Een optimaal economisch beleid voor België en zijn gewesten, volgens de multipele effectiviteitsanalyse*, op. cit., 7-9.

[3] These data are brought together in: W.K.Brauers, *Het Bruto Regionale Product van Vlaanderen, Wallonië en Brussel,* op.cit.

[4] In the mean time the Belgian government is far behind on this schedule.

CHAPTER 3

AN APPLICATION FOR ARMS PROCUREMENT

Belgium has mainly to import its heavy military equipment. Sometimes the contractor or his national authority will offer economic compensations. *Economic Compensations* may have the form of co-production, of compensating purchases from the same economic sector or from other economic sectors[1].

The example of the fighter planes was taken, as it is a domain where prices are more or less known and comparable. Demonstrating how to select a fighter plane does not mean that the author takes a position on this problem[2].

The last delivery and installation of Mid-life Update (MLU) kits for the modernization of the existing F16 would take until the middle of 2003. In order to be effective the planes have to be operational at least for another ten, twelve years. Nevertheless, the lobbyists are already in full action. Will the planes that are offered now not be outdated in 2015? A new generation is already foreseen for 2020. Moreover, it is not sure that Belgium will buy fighter planes in the future. On the one side there is the exponential rise in prices, deadly for a small country, and on the other side a European Defense Community may take shape before that time.

3.1 THE APPLICATION OF THE FULL-MULTIPLICATIVE FORM

The procedure is followed as was described in the previous chapters[3].

1) The Filtering Stage

From the beginning the following planes were withdrawn in a qualitative filtering stage: the planes supposed not to be multi-purpose, such as the French Rafale, and the planes considered as too small, such as the Swedish Gripen. In addition, planes of a next generation are not taken into account, such as the Russian Model 1.42 "ATF'ski" and the F22. These planes would be too expensive for Belgium anyway.

2) Domination

No plane is assumed to dominate the other planes for all the objectives. This means that until now all alternatives are incomparable. A way out has to be found to attain results.

3) Solving the problem of the Until Now Incomparable Alternatives

New fighter planes have to supersede the 72 existing F16s. The Ministry of Defense (MOD) would give four points to its first choice, three to its second one etc. This MOD-ranking is an expression of quality of a cardinal scale. In other words, the four points for the Joint Strike Fighter double its quality as compared with the two points for the Eurofighter. This MOD-ranking is entirely hypothetical. The ranking as first candidate of the Joint Strike Fighter is however realistic as this choice was already defended by the Minister of Defense before the Cabinet of Ministers. As Belgium is already a client of Lockheed with the F16, economic compensations for the Joint Strike Fighter are estimated at 50% of its total cost price.

Being already operational with a good record and with a moderate cost price a new update (a Three Quarter-life Update!) of the MLU F16 is ranked second. There are however indications that economic compensations are not obtainable at that moment.

A certain aversion exists against the Eurofighter, though economic compensations for 80% are possible. Due to the political difficulties with the participants, a high risk factor of 0.9 is assigned[4].

The F18 E/F is ranked last, being not immediately deliverable. Economic compensations are here estimated at 30%.

The person-years as an economic indicator include the total of direct work on the economic compensations and the indirect work coming from repercussions in other sectors.

The deficit of the balance of payments is caused by the payments abroad, if not counterbalanced by economic compensations.

Once the cost price of each plane has been estimated, all the necessary calculations can be made as summarized in the following Table 3.1.

Table 3.1 Data for the Procurement of New Fighter Planes by Belgium (a)

Planes	MOD	Man-years	Deficit B/P in million BEF	Price in billion BEF	Risk
EUROFIGHTER	2	77,849	37,123	2.6	0.9
JOINT STRIKE F.	4	29,840	56,916	1.6	0.5
F18 E/F	1	26,963	120,002	2.4	0.3
F16 (updated)	3	1	115,200	1.6	0.1

(a) With the absence of economic compensations under the form of man-years for plane F16, its multiplicative form would end with zero. This nonsense situation is corrected by giving the low symbolic value of one. Nevertheless, it remains a heavy handicap for its optimum.

On this data formula V about attributes and objectives moving in a different direction (see Part 5, Chapter 1.7), is applied for the different alternatives, including:

- maximization of the military choice
- maximization of the employment in person-years through the economic compensations in Belgium
- minimization of the deficit of the Belgian balance of payments through the importation of foreign equipment
- minimization of the cost price in billion Belgian francs for an imported fighter plane
- minimization of the risk linked to the delivery of a fighter plane being, either not conformable to the agreement, or postponed or even cancelled.

The results are summarized in Table 3.2:

Table 3.2 Final Ranking for the Procurement of New Fighter Planes by Belgium

Planes	U'_I	Final ranking	Rank MOD	Rank man-y.	Rank B/P	Rank price	Rank risk
JOINT STRIKE FIGHTER.	2.62140699	1	1	2	2	1	3
EUROFIGHTER	1.79235533	2	3	1	1	4	4
F18 E/F	0.31206656	3	4	3	4	3	2
F16 (updated)	0.00016276	4	2	4	3	2	1

The Joint Strike Fighter as the first choice of the Ministry of Defense is confirmed by the general analysis, which also includes the economic repercussions.

3.2 DOES THERE EXIST A DANGER FOR A REVERSE ORDER?

As was said earlier a reverse order is acceptable being the result of a change in the stakeholders' attitude but not if it happens outside the will of the stakeholders.

a. Result of a Change in the Attitude of the Stakeholders

Here the significance of the α-parameters enters into the picture. Suppose the government favors the alternative with the highest employment contribution, namely Eurofighter with 77,849 person-years. At that moment, the government put, for instance, α_1 = 77,849. This means that a fictive employment of 77,849 is allotted to Eurofighter. At that moment, the U'_i for Eurofighter is doubled, with Eurofighter ranking first instead of JSF. The federal Council of Ministers or the regions have then to bear the responsibility if military effectiveness is lowered. Additionally the federal government, or eventually the regions concerned, has to pay the increase in price, which otherwise the Ministry of Defense would have to pay:

(2.6 billion BEF x 72) - (1.6 billion BEF x 72) = 72 billion BEF. It also occurs very often that through lobbying an alternative candidate obtains reopening of the period for presenting an offer. Suppose it happens and the F16-people revise their position by also offering a 50% economic compensation, which is translated into:

30,198 man-years activity
57,600 m. BEF balance of payments deficit

awarding the F16 (for the second time updated) the first rank with an indicator of 9.83007813. The indicators for the other planes remain unchanged. The outcome is acceptable on the economic side and for the military as they rank the twice-updated F16 second, whereas the risk factor is very low.

b. Result from the Introduction of a New Alternative

Suppose that the two times updated F16 would be considered an entirely new plane. The old ranking is not reversed beside the dominating position of the new alternative, as shown in Table 3.3.

Table 3.3 Introduction of a New Alternative Plane

1	The new alternative plane	9.83007813
2	JOINT STRIKE FIGHTER	2.62140699
3	EUROFIGHTER	1.79235533
4	F18 E/F	0.31206656
5	F16 (updated)	0.00016276

The other stakeholders would not accept that the military would apply an exponent on their attribute. It is also the case vice versa but the other stakeholders would be not opposed if the military proposed different attributes for defense. Defense indeed introduced only a single attribute but the economic decision makers introduced three, besides the technical factor of risk. Nevertheless, the outcome was especially favorable for the military: the number of attributes (objectives) per stakeholder did not play a role. It does not mean that defense could not introduce several attributes. On the contrary, it would be advisable that for each weapon system defense would judge on several criteria to be considered as separate objectives.

3.3 CONCLUSIONS CONCERNING THE CHOICE OF A FIGHTER PLANE IN BELGIUM

For arms procurement the Belgian government asks for a good equilibrium between quality and price. Quality remains however a vague notion. Therefore, several attributes of objectives, desirable from the military side, could characterize the notion of quality. On the cost side, it is not only the budgetary cost that counts, but also the macro-costs, such as repercussions on employment and on the balance of payments. In other words, it concerns a problem of Multiobjective Analysis.

There is No Bias between the Data of the Alternatives[5]

This conforms to the theory which was developed. The different sizes of the attributes have no effect on the outcome. The extreme case of the same response on an attribute by all alternatives would mean that this attribute would have no effect on the final outcome. Consequently, one may say that weak differences among the responses of the alternatives would also have a weak influence on the final outcome. This last point was not relevant here as the differences among the responses of the alternatives on the attributes were significantly large.

Why not compare with or use Several Methods, which is sometimes suggested?[6]

From Table 3.2 it seems clear that an outranking method, comparing the ranks of the several attributes, would choose the Joint Strike Fighter in the first place, but the ranks for the other planes are rather vague. An additive linear model is not applicable as long as weights are not allotted to the several attributes. The use of weights however leads to extreme results with few chances for an in-between alternative. As seen from Table 3.2, for any choice of weights plane F18 E/F will never have the chance to rank first. On the contrary, the other planes may have that possibility, if a strong weight is given to the attribute for which they are the best. In addition, a Reference Point Theory is not operational without scores in a normalization process. In a deterministic way or via a Delphi exercise, the stakeholders have then to allocate scores to the five attributes, either in monetary units or simply in points. Consequently, for all these methods additional information is needed.

NOTES PART FIVE CHAPTER 3

[1] .W.K.Brauers, *Ons leger waakt! Waakt ons leger wel?* (Dutch for: Our Army is awake) de Vries-Brouwers, Antwerp, 1993.
W.K.Brauers, *Multiple Criteria Decision Making and the Input-Output Model for National Defense*, Department of National Defense, Brussels, 1977.
W.K.Brauers, *Systems Analysis, Planning and Decision Models*, Elsevier, Amsterdam - New York, 1976.

[2] The names of the planes, which are cited, are considered as purely illustrative without any special appreciation from the author. It concerns a first and personal impression based on secondary information from different publications and certainly open for re-examination. The analysis, which follows, is very imaginary. It does not express the point of view of the Belgian Department of Defense or of the cabinet of Belgian ministers.

[3] More details are given in: W.K.Brauers, *The Multiplicative Representation for Multiple Objectives Optimization with an Application for Arms Procurement*, Naval Research Logistics, Wiley Periodicals, Vol.49, 2002, 327-340.

[4] Minimization of risk is considered here as a separate objective, an example of *Risk Independency* (see note 11, Part III, Chapter 1).

[5] Background information about the data in Tables 3.2 and 3.3 is as follows.

Prices of the planes are prices franco Belgium, traditionally not including certain items such as the costs of extra spare parts and the cost to stock the spare parts. The price for the EUROFIGHTER is the price already agreed by Germany for a first booking.

The relation between production (which is directly caused by the economic repercussions) and value added is found in a Belgian input-output table, sector aircraft and similar transport equipment. This is also the case for the relation between value added and employment for this sector. In this last ratio, current prices are taken into account.

Indirect repercussions, because of the importation of fighter planes, are found via the internal multiplier on direct repercussions from Belgian experience.

[6] Suggestion from: K.R.MacCrimmon, *Decisionmaking Among Multiple-Attribute Alternatives: A Survey and Consolidated Approach*, Rand Corporation, Santa Monica, RM-4823-ARPA, 1968.

CHAPTER 4

MULTIPLICATIVE REPRESENTATION AND PROJECT PLANNING FOR A DEVELOPING COUNTRY

In project planning several objectives are taken into consideration. Financial, economic, sociological and ecological objectives are quite different and the decision makers being business managers or public servants serving the general interest will have different objectives in their mind. In this way there is a niche in the market for a method which can compare numerous alternative projects making optimum multiple objectives sometimes with different units of measurement. Multiplicative methods are developed for a set of nondominated alternatives taking into consideration upper limits, lower bounds, indifference attitudes and finally ending in a ranking based on a multiobjective nonlinear utility

4.1 A NATIONAL INVESTMENT BOARD IN A DEVELOPING COUNTRY

A case study is brought, for project planning around a National Investment Board from a developing or a new industrial country.

A project has to be considered in two phases: the feasibility and the execution phase. In the *Feasibility Phase,* opportunities are macro- and microeconomic, technical and sociological examined. The necessary feasibility studies are made on market and plant capacity, material inputs, location and site project engineering, manpower and project implementation and evaluation of investment and financing. The following input factors have especially to be considered:

1) market analysis: market size, market growth rate, the firm's share of the market and selling prices;

2) investment cost analysis with service-life, including, e.g., automation progress;

3) operating and fixed costs.

In managerial sciences an increase in cash flows is possible against a more than proportional decrease in scientific research or in new investments, but at the end nobody will accept a further deterioration in research or in new investments. Modern economics indicates limitations to the available means such as the law of diminishing returns, diseconomies of scale at the end, while sometimes the objective could only be reached at infinity after an asymptotic path. It is the case, for instance, for increasing costs until infinity to avoid totally car accidents on roads.

The introduction of pollution abatement in multiobjective optimization may form a problem. More and more increase in national wealth will go together with a more than proportional deterioration of the ecological situation and at the end nobody will be satisfied with the high ecological cost of the operation.

Anyway, the stakeholders have to agree on a set of objectives. The private manager would think on the short run of maximization of his yearly turn over, of penetration in the national and international markets, of balancing his liquidity position, of minimizing his costs and the payment of taxes (minus subsidies). On the long run, he aims to maximize the Net Present Value and the Internal Rate of Return and to minimize the Payback Period. The local government would like to maximize local employment, the revenue from local taxes and the redistributions in its favor from the national government, whereas the national government likes to maximize its tax revenue. Finally, macroeconomics, through the authority of the government, aim at maximization of direct and indirect employment, of the Economic Net Present Value (ENPV), of the increase in Gross Domestic Product and of the surplus preferably in hard currency in the Balance of Payments.

Most of the time, several projects are put into competition with one another. In the *Execution Phase* either one or several projects are realized.

Suppose we go out from a National Investment Board in a developing country or in a new industrial country, like Thailand; facing multiple objectives and operating inside a five-year national development plan, with one possible extension in mind. Suppose the National Investment Board fixes the following objectives on the attributes of the projects[1]:

1. maximization in dollars of Net Present Value (NPV);
2. maximization of the Internal Rate of Return (IRR) in %;
3. minimization in years of the payback period;
4. maximization of the government income as a % of the government stock invested in the project;
5. maximization of employment in a number of jobs;
6. maximization in dollars of Value Added (VA);
7. maximization in dollars of a surplus in hard currency as shown on the successive balances of payments current account;
8. maximization in dollars of a surplus hard currency for investment on the successive balances of payments;
9. minimization of the risk factor on the deterioration of the terms of trade, as a % of VA.

As a counterpoise the government could guarantee special tax cuts, exemption from export controls and free remittance abroad of dividends. Suppose the Government wants to increase employment and to diminish imports. Both objectives go together if a factory is opened in a developing country instead of importing.

The importer, on his side, may face the following alternative conditions:

a) political events: no change in government, change in government, new revolutionary government, new military government, the government offers satisfactory conditions for establishing a factory, expropriation of factories without compensation, control on profit remittances, change in tariffs and quotas, enforcement to 50 percent local shareholding, agreement continuing unaltered, etc.
b) Economic possibilities: a competitor builds a factory. The alternative solutions for import are: 1) to build his own factory or 2) to build a 50 percent local owned factory.

The estimate of probabilities by Delphi kind of techniques or by simple dialogue between several managers of the firm and the setting up of several decision trees could produce the following advices. If satisfactory terms are offered the importer should build a factory with an equal or a minority share of local ownership. If unsatisfactory conditions were proposed, he should continue to import.

It is decided to use a method with two steps: the filtering stage on the one side and the indifference or the ranking stage on the other.

4.2 THE FILTERING STAGE IN PROJECT PLANNING FOR A DEVELOPING COUNTRY

Suppose several alternative projects facing several objectives, whereas these alternatives are characterized by different attributes. At that moment it is possible in a preliminary stage, called the filtering stage to fix thresholds (lower bounds) or ceilings (upper limits) to these attributes, which will eliminate right away less effective candidates. In the case of project planning, these filters may concern for instance:

a positive Net Present Value;
an upper limit on the payback period;
a lower bound for the Internal Rate of Return;
a minimum government subsidy;
a minimum of market penetration respectively on the domestic, and the world markets;
a discounted surplus on the government participation;
a lower bound on direct employment, on the balance of payments, on the investment in imported hard currency.

Suppose the National Investment Board fixes the following lower bounds and upper limits on the attributes for projects with duration of ten years.

1. a positive Net Present Value (NPV);
2. an Internal Rate of Return (IRR) of at least 13%;
3. the payback period has to come before the end of the 9th year;
4. the government delivering 51% of the project's capital asks for a surplus value of at least 50% of its capital stock (discounted). The surplus value consists of taxes, with a given tax structure known in advance, and revenues;
5. a direct (regional) employment has to be created of at least 200 jobs;
6. over the 10 year period a Value Added of at least 2 m. $ has to be guaranteed;
7. over the 10 years a surplus of hard currency on the balance of payments has to be guaranteed of at least 1 m. dollars;
8. over the 10 year period an imported hard currency for investment of at least 1.250 m. $ has to be provided;
9. a risk factor on the deterioration of the terms of trade of less than 50% of VA.

Seven projects are presented which will be called A, B, C, D, E, F and G. Project D does not guarantee the lower bounds and upper limits concerning IRR, government income and payback period. Project E cannot guarantee the direct employment, project F the minimum value added and project G the hard currencies needed. This means that after a filtering stage only projects A, B and C remain. Project A is very capital-intensive, project C is very labor-intensive and project B is situated between these two extremes.

The competing projects which passed the filtering stage are presented either for the indifference stage or for the ranking stage.

4.3 INDIFFERENCE OR RANKING IN PROJECT PLANNING FOR A DEVELOPING COUNTRY

In real life, many stakeholders are indifferent about alternatives if some thresholds or ceilings are respected. Most of the commodities and services, however, are no free commodities or services and in the traditional scarce economy costs take then a decisive position. Therefore in the Indifference Stage, when the alternatives, here projects, pass the filtering stage and the stakeholders are indifferent about the alternatives, the cost side will determine the optimal choice to be made.

In fact, distinction has to be made between effectiveness and efficiency. Effectiveness will strive for the fulfillment of the objectives, efficiency to do it with the lowest costs[2]. In practice, effectiveness and efficiency belong each to a specialty. Overall planners, scientists or technicians from both the private and the public sector will strive for effectiveness, but financial planners and budgeting people more for efficiency. It is in the sphere of effectiveness that indifference may occur.

With a large number of objectives and a large number of stakeholders, the ranking stage is more operational than the indifferent one.

After the available information, none of the remaining projects seems to dominate strongly or weakly for all objectives. Consequently, they form a set of nondominated alternatives (see Table 4.1).

Table 4.1 Reaction of the Several Projects on the Proposed Objectives i (with i = 1, 2, ..., 9)

Attributes	Project A	Project B	Project C	Scores to be assigned (u_i)
Entrepreneurial economics				
1) NPV	1m. $	1.6 m. $	2 m. $	1 per 100 $
2) IRR	14%	17%	17%	100 per interest point more than 13%
3) Payback period (a)	5 years	7 years	9 years	100 per year earlier
Government (b)				
4) Total government income (c) (discounted)	200%	175%	150%	100 per 5%
Macro-economics				
5) Direct employment (regional) (d)	750	1,000	1,200	10 per job created
6) VA (discounted)	20 m. $	13.5m. $	10 m. $	1 per 2,000 $
7) Surplus in hard currency on Balance of Payments current account over 10 years	4.5 m.$	4.0 m. $	3.5 m. $	1 per 1,000 $
8) Imported hard currency to be provided for investment	2.25m. $	1.50m. $	1.25m. $	1 per 1,000 $
9) Risk factor: deterioration of terms of trade	20%	25%	30%	200 per percentage point less than 50%

(a) Projects have to last at least 10 years.
(b) Government participates for 51 % in the capital of the project.
(c) Including taxes with a given tax structure and revenues from the project, with a fine for non-realization.
(d) Have to be recruited during the first and the second year.

In order to normalize the objectives the stakeholders are asked in a Delphi exercise to indicate a score for each of the proposed objectives for instance in the following way.

A Delphi group is composed as follows:

- four representatives of the Government
- three representatives of the National Investment Board;
- four representatives of the employers associations;
- four representatives of the trade unions.

The Delphi meeting is organized in such a way that communication between the group members is directly impossible (anonymous). Anonymity could also be guaranteed if the meeting is replaced by sending questionnaires.

A panel composed of the three members of the National Investment Board makes proposals. In the beginning, the panel asks the question: "do you suggest other important objectives?" The first round will eventually include the new objectives

Next, the panel proposes for instance one, two, three or more score points per 100 dollars for NPV. Each member allocates some points to the proposed objective. The median and the quartiles of the answers are calculated. If the differences between the median and the quartiles are too large, a new round is run, after eventually mentioning extreme positions. New rounds may be organized until convergence in opinion is satisfactory.

The exercise is repeated for each objective. When scores are attributed to all the objectives, the Delphi group is asked if it agrees on the concordance between all the scores. The agreement asks for a two-thirds majority at least. In case of discordance, the exercise is repeated for the score or scores, which are in discordance. Column five of Table 4.1 indicated the final results: scores to be assigned (u_i).

Table 4.1 provided the elements to calculate the final points to be assigned to each project, which is done in Table 4.2.

Table 4.2 Scores obtained per Project

Attributes	Project A	Project B	Project C
1) N.P.V.	10,000	16,000	20,000
2) I.R.R.	100	400	400
3) Payback Period	400	200	0
4) Government income	4,000	3,500	3,000
5) Employment	7,500	10,000	12,000
6) V.A.	10,000	6,750	5,000
7) Hard currency in Balance of Payments	4,500	4,000	3,500
8) Hard currency for investment	2,250	1,500	1,250
9) Deterioration of Terms of Trade (risk factor)	6,000	5,000	4,000
Totals	44,750	47,350	49,150

If the scores are taken as weights, the linear method produces the following ranking:

C P B P A

Project C can be considered as more labor-intensive, A as more capital-intensive, with Project B as a more in-between alternative. As already indicated earlier in a previous chapter, such an in-between alternative will never rank first in a linear approach. Moreover a linearity with the use of weights has to be excluded being in contradiction with the law of diminishing marginal utility. Instead an expected multiobjective utility function has to be assumed, being nonlinear. Therefore two approximations are possible: the indifference method and the reference point theory.

The indifference method intervenes when the stakeholders are indifferent between the projects once the filtering test is passed. At that moment the project is chosen which guarantees the highest discounted Value Added (macro-economic point of view of the National Investment Board). Therefore, an audit will control for instance the projected sales and expenditures, the several depreciation rates, the forecasting of the salaries and of the working capital and the financial planning for the project. Eventually firms could be fined when the forecasted VA is not reached at the end of the project.

If the following VA are reached (cf., Table 4.1) and given that the stakeholders are indifferent on the basis of the other objectives:

Project A 20 m. $
Project B 13.5 m. $
Project C 10 m.$ then: A P B P C

If the risk factor on the deterioration of the terms of trade is taken into consideration, then the following VA are obtained:

Project A: 20 m. $ x 80% = 16 m. $
Project B: 13.5 m. $ x 75% = 10.125 m. $
Project C: 10 m. $ x 70% = 7 m. $ then: A P B P C

For the ranking stage, we first apply the reference point theory with maximal criterion values. Therefore, the co-ordinates of the reference point are determined with the maximal criterion values. In Table 4.3 the deviations from the reference point are given for each project.

Table 4.3 Deviations with Scores from the Reference Point

Attributes	r_i	A	B	C
1) NPV	20,000	10,000	4,000	0
2) IRR	400	300	0	0
3) Payback period	400	0	200	400
4)Government income	4,000	0	500	0
5) Employment	12,000	4,500	2,000	0
6) VA	10,000	0	3,250	5,000
7) Hard currency in Bal. of P.	4,500	0	500	1,000
8) Hard currency for Investment	2,250	0	750	1,000
9) Deterioration of T. of Trade	6,000	0	1,000	2,000
Largest Deviation		10,000	4,000	5,000

Conclusion: B P C P A

Of course, much depends on the social and economic conditions posited by the stakeholders (not even speaking about the political ones), which are translated in the scores. A multiplicative representation without scores shows a different result (see the following Table 4.4).

Table 4.4 The Multiplicative Representation for Project Planning in a Developing Country

Stages	1	2	3	4	5	6	7	8	9
Projects	MAX. NPV	MAX. IRR	3 = 1 x 2	MIN. Payback	5 = 3 : 4	MAX. Gov. Y	7 = 5x6	MAX. Employm.	9 = 7 x 8
A	10	14	140	5	28	200	5600	75	420000
B	16	17	272	7	38.8571429	175	6800	100	680000
C	20	17	340	9	37.7777778	150	5666.66667	120	680000
10	11	12	13	14	15	16	17	18	
MAX. VA	11= 9 x10	MIN. Risk	13= 11:12	MAX. B. of P.	15= 13x14	Investm.	17= 15x16	Result	Projects
200	84000000	20	4200000	45	189000000	225	4.2525E+10	**1**	A
135	91800000	25	3672000	40	146880000	150	2.2032E+10	**2**	B
100	68000000	30	2266666.67	35	79333333.3	125	9916666667	**3**	C

The calculations from Table 4.4 produce the following ranking:

$$A \; P \; B \; P \; C$$

This time the most capital-intensive project, project A, wins. In order to test this outcome, an exponent of 2 is given to employment. Table 4.5 produces the results.

Table 4.5 The Multiplicative Representation with Exponents for Project Planning in a Developing Country

Stages	1	2	3	4	5	6	7	8	9	10
Projects	NPV	NPV²	IRR	4 = 2 x 3	Payback	6 = 4 : 5	Gov. Y	8 = 6 x 7	Employm.	9²
A	10	100	14	1400	5	280	200	5600	7	5625
B	16	256	17	4352	7	621.7142857	175	10880	10	10000
C	20	400	17	6800	9	755.5555556	150	113333.333	12	14400
11	12	13	14	15	16	17	18	19	20	21
11 = 8 x10	VA	VA²	14 = 11 x 13	Risk	16 = 14 /15	B. of P.	18 = 16 x 17	Investm.	20 = 18 x19	Result
315000000	200	40000	1.26E+13	20	6.3E+11	45	2.84E+13	22	6.37875E+1	**A** 1
1088000000	135	18225	1.98288E+13	25	7.93152E+11	40	3.17E+13	15	4.75891E+1	**B** 2
1632000000	100	10000	1.632E+13	30	5.44E+11	35	1.9E+13	12	2.38E+1	**C** 3

Despite the fact that an exponent is given to employment, the capital-intensive project, project A, is still ranked first.

The outcome of the several methods is contradictory this time. After all, the scores of the in-between alternative, project B, is very satisfactory, although the results of project A, has to be taken into consideration.

NOTES PART FIVE CHAPTER 4

1. Of course, most of the time the number of different objectives will be less than the numerous objectives enumerated here. This wide choice of objectives is given however in order to strive for completeness and to show objectives with different units of measurement.

2. This definition of efficiency stems from systems analysis literature. In multiobjective literature the definition is different. A point is then called efficient if its criterion vector is nondominated.

CHAPTER 5

AN OPTIMAL INPUT-OUTPUT STRUCTURE FOR A DEVELOPING COUNTRY SUCH AS TANZANIA

Explaining the title of this research means that: first, input-output structure in a developing country has to be defined. The reasons have to be given why input-output structure in a developing country would be different from that of a developed country. Second, the application of the full-multiplicative form for an optimal Input-Output Structure in a developing country has to be discussed, especially applied for Tanzania.

5.1 SOME CHARACTERISTICS OF INPUT- OUTPUT IN DEVELOPING COUNTRIES

Concerning Aggregation

Before starting to measure input-output, aggregation of individual firms or products is necessary but extremely difficult in a developed country. In Belgium e.g., the chemical industry counts 10,000 different products. In a developing country, such as Tanzania, aggregation is much easier.

Concerning Technical Structure

The input-output structure consists of a matrix for the totality of the economy in which the deliveries are shown in the rows and the resources in the columns. Both deliveries and resources are expressed in quantities or in value, though mostly in value, and are interrelated by ratios that are called "technical coefficients". A limited stability of these technical coefficients is assumed and in this way the matrix is applied for prognosis and planning, also in its inverted form. In the first research of Leontief, the pioneer of the method, stability of the technical coefficients was assumed for the United States over a period of ten years. In other developed countries a similar stability was assumed. The European Union recommends that its members construct a full table each five years.

By updating the input-output table by methods such as RAS or PES an extension is possible for another five years, meaning that the lifetime of the tables of the West-European countries is extended from five to ten years[1]. The basis input-output table consists of an existing table or is constructed e.g., by *Benchmarking* (similar countries) or by engineering data. Suppose, we start from an input-output table for year 1 ($_1X$), which is assumed valuable for five years. The table is updated by PES to ($_{10}X$). The updated table is valuable until the 10th year inclusive, unless a major and sudden economic disaster occurs (such as the first oil crisis 1974 - 75 for the oil importing countries). In that case an "emergency planning" has to be foreseen.

In a developing country the situation is quite different. On the one side, traditional sectors such as agriculture and fisheries could use production methods that are invariable perhaps for centuries. On the other side, the introduction of a new industrial firm may completely and suddenly upset the technical coefficients. The installation of a new car assembly plant, for instance, will completely change the technical coefficients of the production sector of the means of transportation or will even create a new sector. At that moment, the developing country can use engineering data or can do benchmarking on basis of experience in similar countries. The PES method when updating the input-output table of a developing country will take care of these questions (see therefore Part I, 2.5 of this book).

For a developing country, given traditional sectors on the one hand and new sectors specified by benchmarking on the other, the validity of the input- output table may exceed ten years. This period may be different from one developing country to another.

In addition, the structural changes' stage of the PES method opens the way for setting up alternative input-output structures, which could come into competition with one another. Each input-output structure would form the core of a *Scenario*[2]. Economic policy has then to make a choice between the different scenarios. Another innovation, introduced here, consists of the combination of input-output with "multiple objectives utility theory".

Starting from the updated input-output table economic policy may choose between several scenarios after fixing attributes to the objectives.

5.2 ECONOMIC POLICY WITH SEVERAL INPUT-OUTPUT SCENARIOS AS A BASIS IN A DEVELOPING COUNTRY

Tanzania is taken as an example[3]. Suppose that the following objectives are advanced for Tanzania, with the scenarios as alternatives:

- maximization of employment in man-years;
- maximization of Value Added in the national currency. Why is the maximization of productivity not taken into consideration? Productivity, defined as value added over employment, would by definition exclude the previous objectives of value added and employment. In a developed country productivity may play a more dominant role, but in a developing country employment comes first. However, also in a developing country one has to be aware of structural

development, such as economic growth possibilities. Therefore, the *Internal Activity Multipliers* of Input-Output are taken into consideration (for details see Part I, 2.4);

- maximization between the internal activity multipliers of the scenarios;
- maximization of the balance of payments surplus of each scenario;
- due to shortages:
    - minimization of the technical coefficient per scenario of electricity and water
    - minimization of the technical coefficient per scenario of domestic transportation, linked to bad roads.

The following scenarios are foreseen:

Scenario I

Installation of a hydro-electrical power station and a steel factory.

Scenario II

A new policy for agriculture and fisheries:

- the end of incentives to reduce agricultural production through price controls, export taxes, quotas, import subsidies;
- removing protection for non-agricultural products, which increases the balance of payments deficit;
- liberalization of products such as coffee, cacao, cotton and sugar.

Scenario III

Installation of a new sugar factory and a cotton mill.

Scenario IV

Intensified promotion of tourism together with a program for new roads.

The data on these scenarios are assembled in the following numerical illustration on which the full-multiplicative form is applied (Table 1.1). All the objectives are treated in the same way. Neither α-terms (see Part V, 1.3) nor exponents (see Part V, 1.11) are considered.

Table 5.1 Scenarios for an Optimal Input-Output Structure in Tanzania

	1. Man-years MAX,	2. Value Added MAX. (in m.Tsh)	3. Techn. coeff. electricity and water	4. Techn. coeff. transport (domestic)	5. Internal Multiplier MAX,	6. Sectoral Surplus Bal, of Payments MAX. (in m.Tsh)
			MIN. in 0/0000	MIN. in 0/0000	per 1,000	
I/O I	13584	26054	1405	2367	1449	-10247
I/O II	1918702	595480	294	1431	1278	-3559
I/O III	47347	28345	4289	925	1940	28277
I/O IV	165410	75424	1066	1936	1671	143
	OPERATIONS					
	(7)=(1)*(2)	(8)=(7)/(3)	(9)=(8)/(4)	(10)=(9)*(5)	(11)=(10)*(6)	RANK
I/O I	353917536	251898.6021	106.4210402	154204.0872	15.04870569	4
I/O II	1.14255E+12	3886219956	2715737.216	3470712162	975193.0772	3
I/O III	1342050715	312905.2728	338.2759706	656255.3829	18556933463	1
I/O IV	12475883840	11703455.76	6045.17343	10101484.8	1444512327	2

In the Balance of Payments a difficulty springs from showing a deficit instead of a surplus. Indeed scenarios I and II indicate such a deficit. Of course this deficit has to be minimized whereas a surplus is maximized. In other words the same formula is applied as for objectives moving in a different direction. A fraction is used in which the deficit, provided with a positive sign, is placed in the denominator. Table 5.2 summarizes the final results[4].

Table 5.2 Results from the Scenarios for an Optimal Economic Policy in Tanzania

1) New factories	(I/O III)	18556933463
2) Tourism and roads	(I/O IV)	1444512327
3) Reform of agriculture	(I/O II)	975193
4) Heavy industry	(I/O I)	15

The development of new industry by way of setting up a sugar factory or/and a cotton mill ranks first, followed by giving more importance to a program of new roads and of intensified promotion of tourism, whereas reform of agriculture and the setting up of a hydro-electrical power station and a steel factory, seem less significant. Promotion of tourism is handicapped by the fact that a network of new roads is a preceding condition. This new roads program needs more imports, counterbalancing the balance of payments surplus of tourism revenues.

Is it possible that two or more scenarios show the same outcome? Due to the multiplicative system with sometimes outcomes of the order of 10^{12} and mostly with outcomes diverging far away from one another, this was e.g., the case with the outcomes in Table 5.2; the chance to obtain the same result is very unlikely. If exceptionally it would be the case the distance to a reference point is measured.

5.3 LINKAGES BETWEEN THE SCENARIOS AND THE ECONOMIC POLICY OF TANZANIA

The link with the 3-year Plan

The last available input-output table, namely for 1992, is extremely important for control and amelioration of the national accounts of Tanzania which are undoubtedly underestimated[5].

For economic policy the table is outdated. What can be done in that field? From the statistical point of view first of all the national accounts data are revised at least on the basis of the sectoral corrections as found by the 1992 I/O table.

Second, new national accounts are set up on basis of a new census, the long term economic survey and other direct statistical data.

From the planning point of view the 1992 input-output table is updated. The PES-Method is the short cut method, which could be used (for application of the PES method, see: Part I, 2.5).

As was explained earlier the PES-method consists of three stages:

P for Price adjustments

Actual prices are compared with the 1992 prices by statistical price indices, statistical surveys etc.

E for Extrapolations of Quantities

Extrapolations are made on basis of the national accounts or directly from the census, the economic surveys or eventually from sampling.

S for Structural Changes

As was said earlier, the traditional sectors in developing countries are quite stable in technological structure over a long period. This assumption is made for price and quantity ratios. We may for instance assume that a new I/O table has to be made after 15 years. It is not true concerning the structural changes.

If important changes occurred in some important sectors, then these sectors are separately studied in monographs. The results are then incorporated in the I/O table. If the new sugar factory and the cotton mill of scenario III are built their technological structure is incorporated in the I/O table on basis of engineering data, bench marking etc.

The economic policy in the Plan

The following steps are taken:

1) different scenarios, which are assumed as possible through policy measures, international assistance and investments, are written by the Government;
2) a steering group equally composed of government representatives and external experts will determine the objectives to be maximized or minimized;
3) the *Full-Multiplicative Form* is applied;
4) the ranking of scenarios, which is obtained in that way, is presented to the Government to be used for the Plan;
5) each year the Plan is revised on basis of a revised PES.

For the long term horizon of 2025, which is considered in Tanzania, this procedure is not valuable. A development plan for such a long period is impossible for any country. Rather demographic studies and technological forecasting are useful in this context (cf. Part I, 2.7).

5.4 CONCLUSIONS

Too many input-output applications are only restricted to statistical work. The application for economic policy has to be given more attention. In this application not only economic policy is taken into consideration but also an optimal approach.

Scenario writing in combination with input-output, which forms a part of this model, is not new. The link, however, with a multi-objective utility theory is new, whereby preference is given to the full-multiplicative form.

Special care is taken of developing countries where more than elsewhere attention has to be given to benchmarking, engineering data and structural changes. The final outcome resulted in a ranking of several alternative input-output structures under the form of several alternative scenarios.

The multiplicative representation discusses the use of utility functions with multiple objectives in the context of ranking and/or selecting alternative policies, actions, solutions or projects, simply called "alternatives". The introduction of the notion "attribute" is a necessity in order to represent the measurement of an objective in an alternative.

The advantages and disadvantages of working with a linear version of the utility function, that attaches weights to the multiple objectives, were evaluated. The conclusion was that setting of weights leads to extreme results with few chances for an in-between alternative. Even multilinear forms will show the same disadvantages.

Nonlinear multiobjective theories will ask for normalization, which, as contrasted to weights, is nonadditive. A multiplicative version is defended that even does not require normalization.

Importance to one of the objectives can be given by adding a term in the multiplicative form or by allotting an exponent to an attribute. Introducing the importance notion could mean the introduction of subjectivity. How can subjectivity be avoided as much as possible? Therefore, the use of Delphi or of a heuristic approach is suggested. Delphi sets up a group of especially knowledgeable individuals representing all the stakeholders. A heuristic approach would bring a direct and common consent of all the stakeholders or at least with the strong conviction that all stakeholders will agree on the level of importance of an objective.

The multiplicative representation is only applicable in discrete cases and not for alternatives that are automatically generated during the process, the continuous case.

Unlike the added form, which produces cardinal results, the multiplicative representation produces an *Ordinal Scale*, not a great problem, as the stakeholders are mostly involved in a problem of choice. Nevertheless, significant differences in the dimensionless indicators, obtained after the multiplicative process, will shed some light on the importance, though relative, of the alternatives.

NOTES PART FIVE CHAPTER 5

[1] For more details see: Part I, 2.4 of this book.

[2.] The combination of input-output and scenario writing is not new; see e.g. W.K.Brauers, *Prévisions économiques à l'aide de la méthode entrées-sorties"*, Economica, Paris, 1995, 76.

[3] As any updating is absent the input-output table for 1992 is still used: *Input-Output Table of Tanzania for 1992*, Results of the Input-Output Construction Project supported by the Swedish International Development Cooperation Agency through Statistics Sweden. Research Papers and Reports N° 1 (compiled by S.M.Kapunda, J.M.Komba, R.Staeglin), Presidents' Office, Planning Commission, National Bureau of Statistics, Dar Es Salam, December 1999.

[4] Microsoft Excel was used for the calculations.

[5] R.Stäglin, *Policy-oriented Applications of the new Input-output Table for Tanzania*, paper presented at the Regional Input-Output Conference, Arusha, 12-17 September 1999.

PART SIX

RATIO ANALYSIS A SEPARATE METHOD FOR MULTIPLE OBJECTIVES OPTIMIZATION?

CHAPTER 1

MECHANICAL METHODS AND MULTIOBJECTIVE OPTIMIZATION

Several ways are open to determine weights in a weighting method and the scores for normalization in a distance method.

In the deterministic approach the decision maker fixes the scores in an authoritarian way without taking into consideration the opinions of the other stakeholders.

In the heuristic approach all stakeholders agree from the beginning about weights or scores.

Delphi on the other hand tries to reach all possible stakeholders represented through especially knowledgeable people. Additionally the method is based on anonymity and on numerous feedbacks in order to reach convergence as much as possible. In this way, the method strives after objectivity. Delphi, however, is sometimes a costly exercise.

Finally, mechanical methods, such as *Ratio Analysis*, can be developed to determine weights and scores.

1.1 RATIO ANALYSIS A NEW APPROACH?

The mechanical approach takes then the form of ratios with a common denominator. However, ratio analysis seems to take a separate life. If weights and scores are determined with the previous methods, ratio analysis can be considered as a control method on the other methods, or even as a separate method for optimization with several objectives. This part of the book will develop this line of thought.

Several ratios for normalization can be thought of. We choose some of them, which could be considered as representative.

1.2 THE CHOICE BETWEEN DIFFERENT RATIO SYSTEMS

One of the Alternatives is taken as a Reference

The attributes of one of the alternatives are used as denominators for the corresponding attributes of all alternatives[1].

For instance, the first alternative forms the base:

$$\frac{x_{ij}}{x_{i1}} \qquad \text{for each } x_{ij}$$

with: $j = 1,2,\ldots,m;$ m number of alternatives
$i = 1,2,\ldots,n:$ n number of attributes

However, each alternative can be used as a base, for instance alternative two:

$$\frac{x_{ij}}{x_{i2}} \quad \text{for each } x_{ij}$$

Of course, the outcome will be different for both choices. Will the ranking be different? We shall wait for the answer after giving some applications.

Another problem arises if one of the attributes is missing in an alternative and this alternative is used as a base. Indeed if in the first example where the first alternative is used as a base and for instance:

$$x_{5,1} = 0, \text{ then: } \frac{x_{5,1}}{x_{5,1}} \text{ and all } \frac{x_{5j}}{x_{5,1}} \text{ are undefined.}$$

Therefore, an alternative has to be chosen as a base with no attribute equal to zero.

The Van Delft and Nijkamp Method of Maximum Value

In the Van Delft and Nijkamp Method of Maximum Value the attributes per alternative are divided by the maximum value, which is observed in one of the alternatives[2]:

$$\frac{x_{ij}}{x_i^+} \quad \text{for each } x_{ij}$$

with: $j = 1,2,\ldots,m;$ m number of alternatives
$i = 1,2,\ldots,n:$ n number of attributes
x_i^+ defined as the maximum x_{ij}.

Later on, examples will be given in order to test the validity of this approach.

The Van Delft and Nijkamp Method of Square Roots

In the second Van Delft and Nijkamp method each attribute per alternative is divided by the square root of the sum of the squares of all given values of that attribute in the alternatives[3]:

$$\frac{x_{ij}}{\sqrt{\sum_{j=1}^{m} x_{ij}^2}} \qquad \text{for each } x_{ij}$$

with: $j = 1,2,\ldots,m$; m number of alternatives
$i = 1,2,\ldots,n$: n number of attributes

Later on, examples will be given in order to test the validity of this approach.

The Voogd Ratios

With the Voogd Ratios, each attribute per alternative is divided by the sum of all given values of that attribute[4].

$$\frac{x_{ij}}{\sum_{j=1}^{m} x_{ij}} \qquad \text{for each } x_{ij}$$

with: $j = 1,2,\ldots,m$; m number of alternatives
$i = 1,2,\ldots,n$: n number of attributes

Examples will be given in order to test the validity of this approach.

1.3 RATIO ANALYSIS APPLIED FOR MULTIOBJECTIVE OPTIMIZATION

For multiobjective optimization, weights and scores were until now determined from outside. On the contrary, ratios are determined from inside (mechanical method). We shall from now on use ratios for this multiobjective optimization.

Ratios will enter the picture either in an *Additive Method with Ratios* and multiple objectives or in a *Reference Point Method with Ratios* and multiple objectives. The Additive Method with Ratios produces a final ranking, a good example of multiobjective optimization. In the Reference Point Method with Ratios the ratios replace the products of scores and responses of the alternatives.

The theory on ratio analysis for multiobjective optimization is illustrated with simulations on some of the applications, which were given in the previous parts of this book.

NOTES PART 6 CHAPTER 1

[1] This mechanical approach is comparable with the formula of Schärlig, which multiplies all these fractions, as in Part V, 5.6, point 1 of this book.
A.Schärlig,. *Décider sur plusieurs critères*, Presses Polytechniques Romandes, Lausanne,1985, 76.

[2] A. Van Delft, P. Nijkamp, *Multi-criteria Analysis and Regional Decision-making*, M. Nijhoff, Leiden, Nl.1977, 38. This ratio approach of Van Delft and Nijkamp is applied and written as a formula for the multiplicative form in Part V, 5.6.

[3] A. Van Delft, P. Nijkamp, *op.cit.*, 37. This ratio approach of Van Delft and Nijkamp is also applied and written as a formula for the multiplicative form in Part V, 5.6.

[4] H. Voogd, *Multicriteria Evaluation for Urban and Regional Planning*, Pion ltd., London,1983, 86-87. This ratio approach of Voogd is also applied and written as a formula for the multiplicative form in Part V, 5.6.

CHAPTER 2

RATIO ANALYSIS FOR A DEPARTMENT STORE

In Part Four Chapter 5 of this book, four projects A, B, C and D were proposed to fulfill a set of objectives. Project D was rejected right away in a filtering process. Project A plans to expand not only in the domestic market, but mainly in foreign countries. Project C enlarges only in the domestic market. Project B expands in the domestic market and abroad and brings a midway solution between A and C.

2.1 ONE OF THE ALTERNATIVES IS TAKEN AS A REFERENCE

In ratio analysis where one of the alternatives is taken as a reference, no attribute of this alternative is allowed to be equal to zero. In project C some of the attributes are missing, indeed. Therefore, only projects A and B are taken as a base. The results are given in Annex 1 of this chapter.

Each time two methods are followed. First, an additive method is used without weights or scores, but each time with A or B as a base.

Second, reference point theory is applied, once with A as a base and once with B as a base.

A or B as a base produces different results, which is quite understandable. The outcome of the ranking is more interesting. In three of the four results an unexpected ranking is obtained, namely:

$$C \; P \; B \; P \; A$$

The reference point theory with project B as a base shows the same outcome as previously found in the reference point method with scores, namely:

$$B \; P \; C \; P \; A$$

Consequently, ratio analysis, with one of the alternatives taken as a reference, produces no univocal results.

2.2 THE VAN DELFT AND NIJKAMP METHOD OF MAXIMUM VALUE

A fundamental problem arises with the method of maximum value (see Annex 1 point 2 of this chapter).

In the method of maximum value, the attributes per alternative are divided by the maximum value, or the minimum value, which are observed in one of the alternatives. The ideal situation for minimization occurs when the objective of zero is attained. This could mean dividing by zero. If at that moment the numerator is not zero the fraction is undefined.

In the department store example the deterioration of the terms of trade reaches zero in alternative C, which makes the entire application impossible. Even if to alternative C a symbolic number of for instance 0.001 is given, the result would be biased for the other alternatives. Indeed alternative A in the minimization process would reach the figure of 40/0.001 or 40,000 and alternative B 30/0.001 or 30,000, which means a bias in favor of alternative C. Moreover, in this way the attribute "terms of trade" would solely determine the final ranking of the alternatives, which is not correct.

In addition the method of maximum value is biased as automatically all co-ordinates of the reference point are equal to 1.

With such fundamental handicaps, the method of maximum value is dropped for good.

2.3 THE VAN DELFT AND NIJKAMP METHOD OF SQUARE ROOTS

In the second Van Delft and Nijkamp method each attribute per alternative is divided by the square root of the sum of the squares of all given values of that attribute in the alternatives.

The method is applied for the department store example. Annex 2 of this chapter gives the details of this exercise. Once again, the simulation is done for an additive form without weights or scores on the one side and the reference point theory on the other.

The additive form without weights or scores produces an amazing ranking:

$$\text{B P A P C}$$

i.e., an in-between alternative like B ranks first. Indeed B is a real in-between alternative, as it never ranks first separately for any of the objectives (see Annex 1, Table 2A1 of this chapter). Such a result would never occur with an additive model with weights.

The reference point theory reverses the second and third positions:

$$\text{B P C P A}$$

The reference point theory with scores, as calculated in part IV, Chapter 5 produced the same results.

2.4 THE VOOGD RATIOS

With the Voogd Ratios, each attribute per alternative is divided by the sum of all given values of that attribute in all the alternatives.

In Annex 3, once again the details are given for an adding method without weights or scores on the one side and for the reference point theory on the other. In both cases the following ranking holds:

$$\text{B P C P A}$$

corresponding with previous results.

As already was mentioned earlier, with the Voogd Ratios each attribute per alternative is divided by the sum of all given values of that attribute. The Second Van Delft and Nijkamp Method is a bit more sophisticated than the Voogd Method. Indeed, each attribute per alternative is divided by the square root of the sum of the squares of all given values of that attribute.

Both approaches seem to be good control instruments for the Reference Point Theory. Indeed, calculating the deviations from the reference point, both approaches bring the same result for the department store application as the original version of the reference point. On the other hand, both approaches seem to open new perspectives for an additive method without weights or scores.

In the next chapter, these remarks will be tested on another example, this time from the side of the multiplicative representation.

CHAPTER 2 - ANNEX 1

SOME METHODS OF RATIO ANALYSIS APPLIED FOR A DEPARTMENT STORE

In Part IV, Chapter 5, reference point theory solved the problem of marketing with multiple objectives for consumer goods in a department store. In this annex, ratio analysis controls the results. In addition, is there an alternative theory for multiobjective optimization developed?

1. One of the Alternatives is taken as a Reference

1. In three cases out of four, project C ranks first, solely caused by the penetration in the domestic market. Indeed the influence of this penetration is 20 times higher for C than for A. In this way, the outcome is biased. In Part IV, 5.3, C only ranked second in the Reference Point Theory and even last in the linear approach.

2. If project C forms the base, the results are undefined. Indeed in project C the following objectives are missing, i.e., zero:

penetration in another industrialized country
penetration in Southern Europe
penetration in Eastern Europe.

Due to the absence of international trade, there is no deterioration of the terms of trade possible. This means that with C as the base, the denominator for the deterioration of the terms of trade will be zero. As there is deterioration of the terms of trade for A and B, the corresponding fractions are undefined. Ipso facto, the whole exercise is undefined.

3. If project B forms the base, the results are different from those of the A basis, which is quite understandable.

In addition, the ranking is different after the choice of the base for the Reference Point Theory.

In conclusion, the ratio analysis method "one of the alternatives is taken as a reference" is illogical on several points.

Objectives for a department store

1) Maximization of *Net Present Value* (NPV) in million $
2) Maximization of *Internal Rate of Return* (IRR) in interest percentage

3) Minimization of the *Payback Period*, in years before the end of the year
4) Minimization of taxes and subsidies (inclusive para-fiscal), as a percentage of Value Added
5.1) Maximization of domestic penetration, as a percentage of turn over
5.2) Maximization of penetration in another industrialized country, as a percentage of turn over
5.3) Maximization of penetration in Southern Europe, as a percentage of turn over
5.4) Maximization of penetration in Eastern Europe, as a percentage of turn over
6) Minimization of the deterioration of terms of trade as a percentage of NPV
7) Maximization of national direct employment, in number of jobs
8) Maximization of national indirect and secondary employment, in number of jobs
9) Maximization of National Gross Value Added, in billion of $
10) Maximization of the positive influence on the national Balance of Payments, in million $.

Projects A, B, C and D are considered, but D is already eliminated in a filtering stage. Table 5.2 of Part Four, shows the data for the several projects.

In Table 2A1 project A is taken as a reference and in Table 2A2 project B forms the base.

Table 2A1 Project A taken as a Common Reference

The data

	1	2	3	4	5.1	5.2	5.3	5.4	6	7	8	9	10
A	2,000	17	5	20	0.5	2	20	49	40	500	250	5.5	5
B	1,000	14	5	25	4	1	15	30	10	1,100	550	11	4
C	800	13	4	30	10	0	0	0	0	2,000	1,000	12.5	3
Total	3,800	44	14	75	14.5	3	35	79	50	3,600	1,800	29	12

Ratios with project A as base

	1	2	3	4	5.1	5.2	5.3	5.4	6	7	8	9	10	**Totals**	**Rank**
A	1	1	1	1	1	1	1	1	1	1	1	1	1	7	**3**
B	0.5	0.823529	1	1.25	8	0.5	0.75	0.612245	0.25	2.2	2.2	2	0.8	15.88577	**2**
C	0.4	0.765	0.8	1.5	20	0	0	0	0	4	4	2.272727	0.6	29.7	**1**

Choice of the co-ordinates of the Reference Point equal to the Maximal Criterion Values

r_i	1	1	0.8	1	20	1	1	1	0	4	4	2.272727	1

Deviations from the Reference Point

	1	2	3	4	5.1	5.2	5.3	5.4	6	7	8	9	10	**Max.**	**Ran**
A	0	0	0.2	0	19	0	0	0	1	3	3	1.272727	0	19	**3**
B	0.5	0.176471	0	0.25	12	0.5	0.25	0.387755	0.25	1.8	1.8	0.272727	0.2	12	**2**
C	0.6	0.235	0	0.5	0	1	1	1	0	0	0	0	0.4	1	**1**

Table 2A2 Project B taken as a Common Reference

The Data

	1	2	3	4.	5 1	5.2	5.3	5.4	6	7	8.	9.	10.
A	2,000	17	5	20	0.5	2	20	49	40	500	250	5.5	5
B	1,000	14	5	25	4	1	15	30	10	1,100	550	11	4
C	800	13	4	30	10	0	0	0	0	2,000	1,000	12.5	3
Totals	3,800	44	14	75	14.5	3	35	79	50	3,600	1,800	29	12

Ratios with Project B as Base

	1	2	3	4	5.1	5.2	5.3	5.4	6	7	8	9	10	**Totals**	**Rank**
A	2	1.2143	1	0.8	0.125	0.1333	1.3333	1.6333	4	0.45455	0.45455	0.5	1.25	6	**3**
B	1	1	1	1	1	1	1	1	1	1	1	1	1	7	**2**
C	0.8	0.929	0.8	1.2	25	0	0	0	0	1.81818	1.81818	1.13636	0.75	7.75	**1**

Choice of the Co-Ordinates of the Reference Point Equal to the Maximal Criterion Values

r_i	2	1.2143	0.8	0.8	2.5	1	1.3333	1.6333	0	1.8182	1.8182	1.1363	1.25

Deviations from The Reference Point

	1	2	3	4	5.1	5.2	5.3	5.4	6	7	8	9	10	**Max.**	**Rank**
A	0	0	0.2	0	2.375	0.8667	0	0	4	1.3636	1.363636	0.6364	0	4	**3**
B	1	0.2143	0.2	0.2	1.5	0	0.333333	0.6333	1	0.8182	0.8182	0.1364	0.25	1.5	**1**
C	1.2	0.2.86	0	0.4	0	1	1.333333	1.6333	0	0	0	0	0.5	1.6333	**2**

2. The Van Delft and Nijkamp Method of Maximum Value

The Van Delft and Nijkamp Method of Maximum Value seems to be biased. We mention for instance.

1) Automatically all co-ordinates of the reference point are equal to 1. Indeed, the maximal criterion values are either the maximum value divided by itself or the minimum value divided by itself.

2) If the denominator is zero but the numerator not, the fraction is undefined. Already in the previous method this situation would have occurred, if project C had formed the base. Therefore project C was not taken as a base, but the possibility remained to use either A or B as the base.

In the method of maximum value, however, the whole operation will become undefined. Indeed, in the data for the department store, as there is no international trade in Project C, no deterioration of the Terms of Trade is possible (see the data in Tables 2A1 or 2A2). In other words in the minimization process the denominator is zero, the lowest minimum value. With a deterioration of the terms of trade of 40% of NPV in Project A and of 10% in Project B as numerators in the fractions, the ratios for A and B become undefined. Automatically, the whole exercise becomes undefined.

Therefore, it is better to abandon for good the Van Delft and Nijkamp Method of Maximum Value. Instead, the other methods are preferred.

CHAPTER 2 - ANNEX 2

THE VOOGD RATIOS AND THE VAN DELFT AND NIJKAMP METHOD OF SQUARE ROOTS APPLIED FOR A DEPARTMENT STORE

As already mentioned, with the *Voogd Ratios*, each attribute per alternative is divided by the sum of all given values of that attribute.

The *Second Van Delft and Nijkamp Method* is a bit more sophisticated than the Voogd Method. Indeed, each attribute per alternative is divided by the square root of the sum of the squares of all given values of that attribute.

Both methods seem to be good control instruments for the Reference Point Theory. Indeed, calculating the deviations from the reference point, both methods bring the same result for the department store application as the original version of the reference point: *B P C P A*.

Do these methods also bring a new theory for multiobjective optimization? Anyway, both methods put the midway alternative B in the first place, though B never dominates for an attribute.

This means that these Ratio Methods, even for the addition of ratios, can do what is not possible with the Additive Linear Approach: a Midway Alternative can be ranked First.

Table 2A3 shows the Voogd Ratio Method and Table 2A4 the Van Delft and Nijkamp Method.

Table 2A3 The Voogd Ratios

The data

	1	2	3	4	5.1	5.2	5.3	5.4	6	7	8	9	10
A	2,000	17	5	20	0.5	2	20	49	40	500	250	5.5	5
B	1,000	14	5	25	4	1	15	30	10	1,100	550	11	4
C	800	13	4	30	10	0	0	0	0	2,000	1,000	12.5	3
Totals	3,800	44	14	75	14.5	3	35	79	50	3,600	1,800	29	12

Each attribute divided by the sum of all values of that attribute

														Totals	**Rank**
A	0.5263	0.3864	0.3571	0.2667	0.0345	0.6667	0.5714	0.6203	0.8	0.1389	0.1389	0.1897	0.417	2.2658	**3**
B	0.2632	0.3182	0.3571	0.3333	0.2759	0.3333	0.4286	0.3797	0.2	0.3056	0.3056	0.3793	0.3333	2.4321	**1**
C	0.2105	0.2955	0.2857	0.4	0.6897	0	0	0	0	0.5556	0.5556	0.4310	0.25	2.3021	**2**

Choice of the co-ordinates of the reference point

r_i	0.526	0.3864	0.2857	0.2667	0.6897	0.6667	0.5714	0.6203	0	0.5556	0.5556	0.4310	0.417

Deviations from the reference point

														Max.	**Rank**
A	0	0	0.0714	0	0.6552	0	0	0	0.8	0.4167	0.4167	0.2414	0	0.8	**3**
B	0.263	0.0682	0.0714	0.0667	0.4138	0.3333	0.1429	0.2405	0.2	0.2500	0.2500	0.0517	0.0833	0.4138	**1**
C	0.316	0.0909	0	0.1333	0	0.6667	0.5714	0.6203	0	0	0	0	0.1667	0.6667	**2**

Table 2A4 The Van Delft and Nijkamp Method of Squares

Sum of squares and their square roots

	1	2	3	4	5.1	5.2	5.3	5.4	6	7	8	9	10
A	4000000	289	25	400	0.25	4	400	2401	1600	250,000	62500	30.25	25
B	1000000	196	25	625	16	1	225	900	100	1210000	302500	121	16
C	640000	169	16	900	100	0	0	0	0	4000000	1000000	156.25	9
Sum of squares	5640000	654	66	1925	116.25	5	625	3301	1700	5,460,000	1365000	307.5	50
Square roots	2374.868	25.5734	8.1240	43.8748	10.7819	2.2361	25	57.4543	41.2311	2336.6643	1168.3321	17.5357	7.0711

Attributes divided by their square roots

														Totals	**Rank**
A	0.842	0.6648	0.615	0.4558	0.0464	0.8944	0.8	0.8529	0.9701	0.2140	0.2140	0.3137	0.7071	3.508	**2**
B	0.421	0.5474	0.615	0.5698	0.3710	0.4472	0.6	0.5222	0.2425	0.4708	0.4708	0.6279	0.5657	3.616	**1**
C	0.337	0.5083	0.492	0.6838	0.9275	0	0	0	0	0.8559	0.855921	0.7128	0.4243	3.445	**3**

Co-ordinates of the reference point equal to the maximal criterion values

r_i	0.842	0.6648	0.492	0.4558	0.9275	0.8944	0.8	0.8529	0	0.8559	0.8559	0.7128	0.7071

Deviations from the reference point

														Max.	**Rank**
A	0	0	0.12309	0	0.88110	0	0	0	0.97014	0.64194	0.64194	0.39919	0	0.97014	**3**
B	0.421	0.11731	0.12309	0.11396	0.55649	0.44721	0.2	0.3307	0.24254	0.38516	0.38516	0.08554	0.14142	0.55649	**1**
C	0.505	0.15641	0	0.22792	0	0.89443	0.8	0.85285	0	0	0	0	0.28284	0.89443	**2**

CHAPTER 3

RATIO ANALYSIS FOR ARMS PROCUREMENT

In Part V, chapter 3, the problem of arms procurement in Belgium was discussed already. Belgium has mainly to import its heavy military equipment, but tries to do it in the most effective way. The example given concerned the procurement of new fighter planes.

3.1 THE CHOICE OF FIGHTER PLANES

In buying new fighter planes the optimization of the economic repercussions for Belgium are summarized as follows:

- maximization of employment in man-years thanks to the economic compensations in Belgium;
- minimization of the deficit of the Belgian balance of payments caused by the importation of foreign equipment;
- minimization of the cost price in billion Belgian francs for an imported fighter plane.

Table 3.1 Data for the Procurement of New Fighter Planes in Belgium

Fighter planes	MOD	Man-years	Deficit B/P in m. BF	Price in billion BF	Risk
EUROFIGHTER	2	77,849	37,123	2.6	0.9
JOINT STRIKE F.	4	29,840	56,916	1.6	0.5
F18 E/F	1	26,963	120,002	2.4	0.3
F16(updated)	3	1	115,200	1.6	0.1

For this problem the *Multiplicative Form* was used, whereby the co-ordinates were considered as of a dimensionless kind. At the same time, a heuristic approach was playing in the following way. The other stakeholders would not accept that the military would apply an exponent on its attribute. It is also the case vice versa. The other stakeholders, however, would not be opposed if the military proposed several attributes. Indeed, if the risk factor is considered as a technical matter, defense introduced only a single attribute but the economic decision makers introduced three. On these grounds, the following ranking was the result:

1) Joint Strike Fighter
2) Eurofighter
3) F18 E/F
4) F16 (updated)

Notwithstanding that defense only proposed a single attribute, the outcome was especially favorable for the military; the number of attributes (objectives) per stakeholder did not play a role. It does not mean that defense could not introduce several attributes. On the contrary, it would be advisable that for each weapon system defense would judge on several criteria, considered as objectives.

Why not compare with or use several methods, as is suggested sometimes? For instance, comparison with an *Additive Multiobjective Approach with Weights* is a possibility. The use of *Weights*, however, leads to extreme results with few chances for an in-between alternative. For any choice of weights, plane F18 E/F will never have a chance to rank first. On the contrary, the other planes may have that possibility, if the attributes, for which they are the best, receive a strong weight. Even plane F16 (updated), classified last, can rank first due to its low risk profile. If the highest weight is given to the Ministry of Defense (for instance 0.50 and 0.125 for the other attributes) JSF ranks first. This is also the case if employment in man-years receives the highest weight (here namely 0.5 against 0.125 for the others). If the deficit of the Belgian balance of payments, caused by the importation of foreign equipment, receives the highest weight (for instance 0.60 and 0.10 for the other attributes), Eurofighter ranks first. If the risk attribute receives the highest weight, F16 ranks first. Even if weights are given to the F18 E/F, which favors it the most, for instance 0.6 to the risk profile; 0.2 to the cost price and 0.18 to the man-years, the F18 E/F will still end in third place.

For the *Reference Point Theory*, based on the Minkowski metric, *Scores* will make the link between the different measures of the co-ordinates of the attributes. Anyhow, in-between alternatives may have a chance here, which is not the case with weights.

The mechanical approach takes the form of *Ratios*, fractions with a common denominator. These ratios can be used as weights or scores, but we shall go farther by giving them an independent life.

A choice of ratios was made, which can be considered as representative.

3.2 RATIOS ANALYSIS AS APPLIED FOR ARMS PROCUREMENT

One of the Alternatives is taken as Reference

First, the plane F16 (updated) is taken as a reference. Although the final ranking is the same for the additive multiobjective approach in ratio analysis as for the reference point approach (EUROFIGHTER; JSF; F18 E/F; F16/U), the ranking is biased as it is found that only the employment figures determine the outcome. Even worse, if the reference base is changed for instance to the plane, EUROFIGHTER, the ranking is entirely different: 1) JSF 2) EUROFIGHTER 3) F16/U 4) F18 E/F.

As the results of this ratio depend on the choice of the reference, this ratio is not suitable. Therefore, the details of the calculations are dropped. Even for the department store simulation, these ratios, with different reference bases, were leading to contradictory results (see this Part of the book, Chapter 2; 2.1). On the ground of all these negative arguments, this ratio is not worth considering anymore.

The Van Delft and Nijkamp Method of Maximum Value per Attribute

In the department store simulation, the first ratio system of van Delft and Nijkamp, where the attributes per alternative are divided by their maximum value, was evaluated negatively and dropped for good (see this Part of the book, Chapter 2; 2.2). On the ground of the same arguments, we do the same for the choice of fighter planes

However, the Voogd method concerning the division of each attribute by the total of all values of that attribute and the Van Delft and Nijkamp method, whereby the attribute of each alternative is divided by the square root of the sum of the squares of all given values of that attribute, are worthwhile to consider in detail.

The Ratios of Voogd

The Ratios of Voogd form another kind of mechanized system by.dividing each attribute by the sum of all given values of that attribute.

Divided by the sum of each attribute the ratios of the attributes become more equilibrated than in the previous approaches, as is shown in Table 3.2.

Table 3.2 Each Attribute / Sum of all Values of that Attribute for the Procurement of Fighter Planes in Belgium

The data

Planes	MOD	Man-years	Deficit B/P	Price	Risk
EUROF.	2	77849	37123	2.6	0.9
JSF	4	29840	56916	1.6	0.5
F18EF	1	26963	120002	2.4	0.3
F16/U	3	1	115200	1.6	0.1
Totals	10	134653	329241	8.2	1.8

Ratios: each attribute divided by the total of all values of that attribute

Attrib/sum	MOD	Man-years	Deficit B/P	Price	Risk	**Totals**	**Rank**	
EUROF.	0.200	0.578145307	0.112753272	0.317073171	0.5	-0.152	**2**	EUROF.
JSF	0.400	0.221606648	0.172870329	0.195121951	0.277777778	-0.024	**1**	JSF
F18EF	0.100	0.200240618	0.36448073	0.292682927	0.166666667	-0.524	**4**	F18/EF
F16/U	0.3	7.4265E-06	0.349895669	0.195121951	0.055555556	-0.30056575	**3**	F16/U

Choice of the co-ordinates of the reference point equal to the maximal criterion values

r_i	0.400	0.578145307	0.112753272	0.195121951	0.055555556

Deviations from the reference point

Planes						**Max.**	**Min. rank**	
EUROF.	0.200	0	0	0.12195122	0.444444444	0.444444444	**3**	EUROF.
JSF	0.000	0.356538659	0.060117057	0	0.222222222	0.356538659	**1**	JSF
F18EF	0.300	0.377904688	0.251727458	0.097560976	0.111111111	0.377904688	**2**	F18/EF
F16/U	0.100	0.57813788	0.237142397	0	0	0.57813788	**4**	F16/U

Table 3.3 Method of Squares applied for the Procurement of Fighter Planes in Belgium

Squares and square roots of attributes

Planes	MOD	Man-years	Deficit B/P	Price	Risk
EUROF.	4.000	6060466801	1378117129	6.76	0.81
JSF	16.000	890425600	3239431056	2.56	0.25
F18EF	1.000	727003369	14400480004	5.76	0.09
F16/U	9	1	13271040000	2.56	0.01
Sum of squares	30.000	7677895771	32289068189	17.64	1.16
Square roots	5.477	87623.60282	179691.5919	4.2	1.077032961

Attributes divided by the square roots

Planes						Totals	Rank	Planes
EUROF.	0.365	0.888447832	0.206592861	0.619047619	0.835629022	-0.408	**2**	EUROF.
JSF	0.730	0.340547513	0.3167427	0.380952381	0.464238345	-0.091	**1**	JSF
F18EF	0.183	0.307713894	0.66782201	0.571428571	0.278543007	-1.028	**4**	F18EF
F16/U	0.548	1.14125E-05	0.641098444	0.380952381	0.092847669	-0.567	**3**	F16/U

Co-ordinates of the reference point equal to the maximal criterion values

Ref.Point	0.730	0.888447832	0.206592861	0.380952381	0.092847669

Deviations from the reference point

Planes						Max.	Min Rank	Planes
EUROF.	0.365	0	0	0.238095238	0.742781353	0.742781353	**3**	EUROF.
JSF	0.000	0.54790032	0.110149839	0	0.371390676	0.54790032	**1**	JSF
F18EF	0.548	0.580733939	0.461229149	0.19047619	0.185695338	0.580733939	**2**	F18EF
F16/U	0.183	0.88843642	0.434505584	0	0	0.88843642	**4**	F16/U

The Van Delft and Nijkamp Method of Square Roots

In this method of square roots the attribute of each alternative is divided by the square root of the sum of the squares of all given values of that attribute.

The computation of this ratio, as is shown in previous Table 3.3, is more complicated than for the previous ratios. The final ranks, as well for the additive multiobjective approach as for the reference point, are exactly the same as in the ratio case of Voogd, where the complications with squares and roots are not encountered.

3.3 CONCLUSIONS ON ARMS PROCUREMENT

The conclusions on arms procurement are summarized in the following table 3.4.

Table 3.4 Comparison of Different Methods for Multioptimization in the Example of Arms Procurement

Additive Linear Methods with Weights	no weights available
Reference Point Method	no scores available
Indifference Method (on basis of price)	F16/U **I** JSF **P** F18E/F **P** EUROF.
Multiplicative Representation (see Part V, 3.2)	JSF **P** EUROF. **P**F18 E/F **P** F16/U
Additive Method with Voogd Ratios	JSF **P** EUROF. **P** F16/U **P** F18E/F
Additive Method with Van Delft and Nijkamp Ratios	JSF **P** EUROF. **P** F16/U **P** F18E/F
Reference Point Method with Voogd Ratios	JSF **P** F18E/F **P** EUROF. **P** F16/U
Reference Point Method with Van Delft and Nijkamp Ratios	JSF **P** F18E/F **P** EUROF. **P** F16/U

The indifference method has no sense in this context. Indeed, the different stakeholders were not indifferent.

A multiplicative representation with exponents is not mentioned, as the stakeholders would be opposed to it, but not against assigning more attributes to the military (see, p.264). In that way, defense would gain in importance. Indeed, defense only introduced a single attribute for arms procurement, instead of specifying e.g., several qualifying attributes for this arms procurement. Nevertheless, the outcome would not have been changed, as also defense preferred the Joint Strike Fighter in the first place.

In the additive methods with ratios, all ratios are smaller or equal to one. In addition, attributes are treated on an equal basis. In order to avoid all the disadvantages of weights and after the example of arms procurement; the group of stakeholders could agree to allocate more qualifying attributes to one or more participating partners.

In the reference point methods with ratios, ratios substitute the products of scores and responses of the alternatives from traditional Reference Point Theory.

CHAPTER 4

RATIO ANALYSIS AND PROJECT PLANNING FOR A DEVELOPING COUNTRY

In chapter 4 of Part V of this book, different projects for project planning for a developing country were taken into consideration. A case study was brought, for project planning around a National Investment Board from a developing or a new industrial country.

4.1 THE POLICY OF A NATIONAL INVESTMENT BOARD IN A DEVELOPING COUNTRY

The National Investment Board fixed the following objectives on the attributes of the projects:

1. maximization in dollars of Net Present Value (NPV);
2. maximization of the Internal Rate of Return (IRR) in %;
3. minimization in years of the payback period;
4. maximization of the government income as a % of the government stock invested in the project;
5. maximization of employment in a number of jobs;
6. maximization in dollars of Value Added (VA);
7. maximization in dollars of a surplus in hard currency as shown on the successive balances of payments current account;
8. maximization in dollars of a surplus hard currency for investment on the successive balances of payments;
9. minimization of the risk factor on the deterioration of the terms of trade, as a % of VA.

As a counterbalance the government could guarantee special tax cuts, exemption from export controls and free remittance abroad of dividends. Suppose the Government wants to increase employment and to diminish imports. Both objectives go together if a factory is opened in a developing country instead of importing.

The National Investment Board fixed the following lower bounds and upper limits on the attributes for projects with duration of ten years.

1. a positive Net Present Value (NPV);
2. an Internal Rate of Return (IRR) of at least 13%;
3. the payback period has to come before the end of the 9th year;
4. the government delivering 51% of the project's capital asks for a surplus value of at least 50% of its capital stock (discounted). The surplus value consists of taxes, with a given tax structure known in advance, and revenues;
5. a direct (regional) employment has to be created of at least 200 jobs;
6. over the 10 year period a Value Added of at least 2 m. $ has to be guaranteed;
7. over the 10 years a surplus of hard currency on the balance of payments has to be guaranteed of at least 1 m. $;
8. over the 10 year period an imported hard currency for investment of at least 1.250 m. $ has to be provided;
9. a risk factor on the deterioration of the terms of trade of less than 50% of VA.

Seven projects were presented: A, B, C, D, E, F and G. Project D did not guarantee the lower bounds and upper limits concerning IRR, government income and payback period. Project E could not guarantee the direct employment, project F the minimum value added and project G the hard currencies needed. This means that after a filtering stage only projects A, B and C remained. Project A is very capital-intensive, project C is very labor-intensive and project B is situated between these two extremes.

4.2 THE INDIFFERENCE OR THE RANKING STAGE WITH RATIO ANALYSIS FOR A DEVELOPING COUNTRY

After the available information none of the remaining projects seemed to dominate strongly or weakly for all objectives together. Consequently they form a set of nondominated alternatives.

In order to normalize the objectives the stakeholders were asked in a Delphi exercise to indicate a score for each of the proposed objectives. Next the number of scores obtained per project was calculated.

If the scores were taken as weights, the *Additive Linear Method with Weights* produced the following ranking (see Part V, Table 4.1):

C P B P A

Project C can be considered as more labor-intensive, A as more capital-intensive, with Project B as a more in-between alternative. As already indicated earlier in a previous chapter, such an in-between alternative will never rank first in a linear approach. Instead, an expected multiobjective utility function has to be assumed, being nonlinear. Therefore, two approximations are possible: the Indifference Method and the Reference Point Theory.

The *Indifference Method* intervenes when the decision makers are indifferent between the projects once the filtering test is passed. At that moment the project is chosen which guarantees the highest discounted Value Added (macro-economic point of view of the National Investment Board).

Given that the decision makers are indifferent on the basis of the other objectives (see Part V, 4.3):

A P B P C

Applying the *Reference Point Theory* with maximal criterion values, deviations from the reference point were obtained, with the following ranking as a result (see Part V, Table 4.2):

B P C P A

Without scores, without (see Part V Table 4.4) or with exponents (Part V, Table 4.5), the *Multiplicative Representation* shows a different result:

A P B P C.

What is found with ratio analysis? The next tables show the results, respectively for the Voogd ratios in Table 4.1 and the Van Delft and Nijkamp Method of Square Roots in Table 4.2.

Table 4.1 The Ratios of Voogd for Project Analysis in a Developing Country

The Data

Projects	1. NPV (in m.$) MAX.	2. IIR (in %) MAX.	3. Payback Period (in years) MIN.	4. Govern. Income (in%) MAX.	5. Direct Employm. (in jobs) MAX.	6 . V.A. (in m.$) MAX.	7. Balance of Paym.ents (m.$) MAX.	8. Risk T. of Trade (in%) MIN.	9. Import currency for I (m.$) MAX.
Project A	1	14	5	200	750	20	4.5	20	2.25
Project B	1.6	17	7	175	1,000	13.5	4	25	1.5
Project C	2	17	9	150	1,200	10	3.5	30	1.25
SUM	4.6	48	21	525	2,950	43.5	12	75	5

Each Attribute divided by the Sum of all values of that Attribute

Projects										TOTAL	RANK
Project A	0.2174	0.2917	0.2381	0.3810	0.2542373	0.46	0.375	0.2666667	0.4500	1.9243	1
Project B	0.3478	0.3542	0.3333	0.3333	0.3389831	0.31	0.3333333	0.3333333	0.3000	1.6513	2
Project C	0.4348	0.3542	0.4286	0.2857	0.4067797	0.23	0.2916667	0.4	0.2500	1.4244	3

Choice of the Co-Ordinates of the Reference Point

r_i	0.4348	0.3542	0.2381	0.3810	0.4067797	0.46	0.375	0.2666667	0.4500

Deviations from the Reference Point

Projects										MAX.	MIN. RANK
Project A	0.217	0.063	0	0	0.1525424	0	0	0	0	0.217	2
Project B	0.087	0	0.0952	0.0476	0.0677966	0.149	0.0416667	0.0666667	0.1500	0.1500	1
Project C	0	0	0.1905	0.0952	0	0.23	0.0833333	0.1333333	0.2000	0.22989	3

Table 4.2 The Van Delft and Nijkamp Method of Square Roots for Project Analysis in a Developing Country

Projects	1. NPV (in m.$). MAX.	2. IRR (in %). MAX.	3. Payback Period (in years) MIN.	4. Govern. Income (in percent.) MAX.	5. Direct Employm (jobs) MAX.	6. VA (in m. $). MAX.	7. Balance of Paym. (in m.$). MAX	8. Risk Terms of Tr. (in percent.) MIN.	9 Inv.Imp. Currency (in m.$). MAX
A	1	14	5	200	750	20	4.5	20	2.25
B	1.6	17	7	175	1,000	13.5	4	25	1.5
C	2	17	9	150	1,200	10	3.5	30	1.25
Totals	4.6	48	21	525	2,950	43.5	12	75	5

Totals of Squares and their Square Roots

Projects									
A	1	196	25	40000	562500	400	20.25	400	5.0625
B	2.56	289	49	30625	1000000	182.25	16	625	2.25
C	4	289	81	22500	1440000	100	12.25	900	1.5625
Sum Squares	7.56	774	155	93125	3002500	682.25	48.5	8.875	8.875
Square roots	2.74955	27.82086	12.44990	305.16389	1732.772345	26.11992	6.96419	43.87482	2.97909

Attributes divided by their Square Roots

										SUM	RANK
A	0.36370	0.50322	0.40161	0.65539	0.43283	0.76570	0.64616	0.45584	0.75526	3.26481	1
B	0.58191	0.61105	0.56225	0.57346	0.57711	0.51685	0.57437	0.56980	0.50351	2.8062	2
C	0.72739	0.61105	0.72290	0.49154	0.69253	0.38285	0.50257	0.68376	0.41960	2.7538	3

Co-Ordinates of the Reference Point equal to the Maximal Criterion Values

r_i	0.72739	0.61105	0.40161	0.65539	0.69253	0.76570	0.64616	0.45584	0.75526

Deviations from the Reference Point

										MAX.	RANK
A	0.36370	0.10783	0	0	0.25970	0.	0	0	0	0.36370	2
B	0.14548	0	0.16064	0.08192	0.11542	0.24885	0.07180	0.11396	0.25175	0.25175	1
C	0	0	0.32129	0.16385	0	0.38285	0.14359	0.22792	0.33567	0.38285	3

The project planning in a developing country after the example, which was developed here, offers the possibility to compare several different methods for multioptimization. The results, which were obtained earlier for this developing country, are summarized in the following Table 4.3.

Table 4.3 Comparison of several Methods for Multioptimization in an Example of Project Analysis for a Developing Country

Additive Linear Methods with Weights	C P B P A
Indifference Method	A P B P C
Reference Point Method with scores	B P C P A
Multiplicative Representation	A P B P C
Multiplicative Representation with Exponents	A P B P C
Additive Method with Voogd Ratios	A P B P C
Additive Method with Van Delft and Nijkamp Ratios	A P B P C
Reference Point Method with Voogd Ratios	B P A P C
Reference Point Method with Van Delft and Nijkamp Ratios	B P A P C

For project planning in a developing country, the linear method with weights is not taken into consideration given all the negative points, which were raised against it earlier. Anyhow, to assign a first rank to project C is very exceptional.

The indifference method places A in a first rank, which is confirmed by other methods. However, we doubt that for so many objectives and with so many different stakeholders indifference is possible. Contrary to the multiplicative representation and the multiplicative representation with exponents, which also give full priority to project A, the reference point theory with scores prefers the in-between alternative B. Were the scores not well chosen by the Delphi group? Therefore, a control by ratio analysis is interesting. However, the two reference point methods with ratios give also preference to B.

The additive Voogd ratios, as well as the additive Van Delft and Nijkamp ratios give full preference to project A. In the additive method with ratios all attributes are treated equally. Therefore, one may argue to allocate weights to the attributes, now represented by their ratios. At that moment all arguments against weights are taken up again: introduction of subjectivity in the weights, creation of a superobjective, etc. In order to avoid the disadvantages of weights, importance can be given by an increase in the number of attributes related to the objectives. This suggestion took shape into the example on arms procurement in previous chapter.

Of course, inferior objectives and attributes have to be avoided. Moreover, the choice of objectives and attributes has to be the result of a consensus between all the stakeholders involved.

The control function of ratio analysis is very well illustrated in this way.

CHAPTER 5

CONCLUSIONS ON RATIO SYSTEMS

Concerning the mechanical methods different ratios were tested on their reliability.

5.1 CHOICE OF RATIOS

The following conclusions are drawn from these exercises.

1) In the ratio system where the attributes of one alternative were used as denominators, as applied for arms procurement, the ranking was biased as only one objective, namely the employment level, determined the outcome (Chapter 3). If some of the attributes are missing in an alternative, this alternative cannot be used as a base. Even worse, if the reference base was changed the ranking was entirely different (see the application for a department store in Chapter 2).

2) With the ratios of Van Delft and Nijkamp where the attributes per alternative are divided by their maximum value, the result was biased when a zero appears in the denominator of a fraction (example of the department store in 2.2). Automatically, all co-ordinates of the reference point being equal to 1, forms another bias for the method (see 2.2).

3) In the second ratio system of Van Delft and Nijkamp the attribute of each alternative is divided by the square root of the sum of the squares of all given values of that attribute. The computation of this ratio is more complicated than for the other ratios. Moreover, the final ranks, as well as for the additive multiobjective approach as for the reference point, are the same as in the ratio case of Voogd, where the complications with squares and roots are not made. Indeed, in the system of Voogd, each attribute is simply divided by the sum of all values of that attribute.

All these tests seem to suggest that the Voogd Ratio is preferable above the other ones, with the second Van Delft and Nijkamp Ratio as a more complicated version with similar, but perhaps more accurate, results.

Does a unique ratio system exist? Not necessarily. We found a certain similarity between the Voogd Ratio and the second Van Delft and Nijkamp Ratio. It is true that the ratio system, with the attributes of one alternative used as denominators, and the ratios of Van Delft and Nijkamp, with the attributes per alternative divided by their maximum value, were rejected for the tests. Nevertheless they represent entirely other systems than the two cited previously.

By definition in ratio analysis the alternatives have to be known (*Discrete Case*). Ipso facto, a *Continuous Case* has no sense here.

5.2 SIGNIFICANCE OF THE RATIO SYSTEM

Ratios as Weights or Scores

First, ratios as fractions with a common denominator of the attributes can be used to replace weights or scores. However, we see the ratio system in a more independent role.

Ratios as a Control System

MacCrimmon suggested already in 1968 to compare several methods for decision making among multiple-attribute alternatives[1]. In this way the ratio system can be used to test and control all methods, which try to optimize multiple independent objectives. In previous chapters this test occurred on reference point theory, namely on the example of the department store and on the full-multiplicative form with the example of arms procurement.

Is it not possible to go even farther and to propose the ratio system as a new method for optimization of multiple independent objectives?

5.3 THE RATIO SYSTEM AS A NEW METHOD FOR MULTIOBJECTIVE OPTIMIZATION

In order to clarify the issue, the formulas of the several methods are reintroduced.

The Formula of the Additive Linear Method with Weights

The attributes of the objectives are added by using weights. Weights are defined as the importance, which is attached to each of the objectives in a set of mutually exclusive and collectively exhaustive objectives and the sum of the weights, assigned to the elements of this set, has to be one:

$$u(x_j) = \sum_{i=1}^{n} w_i \, . \, x_{ij}$$

with: $u(x_j)$ as total utility for the multiple objectives of alternative j

w_i = weight for attribute i with: $\Sigma w_i = 1$

$i = 1,2,\ldots,n;$ n the number of attributes

x_{ij} = response of alternative j on attribute i

Minimization is translated in a negative sign in the formula.

The linear approach with the use of weights leads to extreme results with few chances for an in-between alternative. In addition, linearity hurts reality[2].

The Min-Max Metric in Reference Point Theory

The Formula for Normalization with Scores

Due to different units of measurement for the attributes of the objectives, normalization has to occur. Normalization means that a score is given to each attribute. Scores, as contrasted with weights, are nonadditive.

$$x_j = [s_1x_{1j},\ s_2x_{2j},\dots,\ s_ix_{ij},\dots\ s_nx_{nj}]$$

with: s_i = the score of attribute i as determined by the normalization process ($i = 1,2,\dots n$, n the number of attributes)
x_{ij} = response of alternative j on attribute i of objective i
x_j = the row vector of utility for alternative j ($j = 1,\dots,m$; m the number of alternatives).

Minima are set over in maxima by allocating the highest scores to the alternatives with the lowest figures of an attribute.

In this way a normalized matrix is obtained:

$$[{}_Nx_{ij}]$$

N indicates the normalization
${}_Nx_{ij}$ = the normalized attribute i of alternative j.

The Formula of the Min-Max Metric

If the following matrix is given:

$$[r_i - {}_Nx_{ij}]$$

with: $i = 1, 2,\dots, n$ as the attributes
$j = 1, 2,\dots, m$ as the alternatives
r_i = the ith co-ordinate of the reference point
${}_Nx_{ij}$ = the normalized attribute i of alternative j,

then this matrix is subject to:

$$\underset{(j)}{Min}\ \{\underset{(i)}{max}\ (r_i - {}_Nx_{ij})\}$$

This distance function forms the basis of Reference Point Theory.

The Formula for the Full-Multiplicative Form

$$U_j = \prod_{i=1}^{n} (\alpha_i + x_{ij})$$

with: $j = 1,2,...,m$; m the number of alternatives
n being the number of attributes and objectives
x_{ij}= response of alternative j on attribute i of objective i
the numbers α are measures of importance.
U_j = overall utility of alternative j. U_j is a dimensionless indicator.

Minimization is treated as a denominator of a fraction with the maximization in the numerator.

A Formula for the Additive Method with Ratios

We tried to set up a similar formula for a general ratio system:

$$y_j = \sum_{i=1}^{i=n} \frac{x_{ij}}{z_{ij}}$$

with: y_j = alternative j valued after the ratio system
$j = 1,2,...,m$; m the number of alternatives
n being the number of attributes and objectives
x_{ij}= response of alternative j on attribute i of objective i

Using the Voogd ratio:

$$z_{ij} = \sum_{j=1}^{j=m} x_{ij}$$

Using the Van Delft and Nijkamp ratio:

$$z_{ij} = \sqrt{\sum_{j=1}^{j=m} x_{ij}^2}$$

All ratios are smaller or equal to one, but contrary to weights their sum can be larger than one in the additive method with ratios.

In the additive methods with ratios all attributes are treated equally. In order to give more importance to an objective, it can be represented by different subobjectives and attributes simultaneously. This was suggested in the example on arms procurement explained in previous chapters.

Of course, inferior objectives and attributes have to be avoided. Moreover, the choice of objectives and attributes has to be the result of a consensus between all the stakeholders involved.

The Reference Point Methods with Ratios

The Voogd and Van Delft and Nijkamp ratios replace the products of scores and responses of the alternatives from traditional Reference Point Theory.

The Ratio System, besides its control function, can become a full-fledged system for multiobjective optimization.

NOTES PART 6 CHAPTER 5

1 K.R.MacCrimmon, *Decisionmaking among Multiple-Attribute Alternatives: A Survey and Consolidated Approach*, Rand Corporation, Santa Monica, RM-4823-ARPA, 1968.

2 "In a very complicated world where many things don't operate in a tidy linear way, you need the best science to tell you whether an action will have the desired consequence" (Sir Robert May, President of the Royal Society, previously head of the Office of Science and Technology of the British Government. Newsweek, 10-9-2000, 76).

CHAPTER 6

GENERAL CONCLUSIONS OF THE BOOK

The optimization of the objectives for a certain domain of the stakeholders society forms the purpose of this book. With stakeholders all persons, interested in an issue, are engaged, which is in contrast with the more limited traditional meaning of decision makers.

Humans have many aspirations, which we translate as "objectives. An old dictum of economics says that human aspirations are numerous, but the corresponding means are limited. If we want to optimize the realization of the aspirations or the objectives in an operational way, we have to work with a limited number of objectives. Is this contradictory with what was said earlier? Happily, some objectives seem to be interrelated in an interdependent or even dependent manner. Therefore, a preliminary search is necessary concerning dependencies and interdependencies of the objectives in order for them to be better understood (see Fig. 6.1).

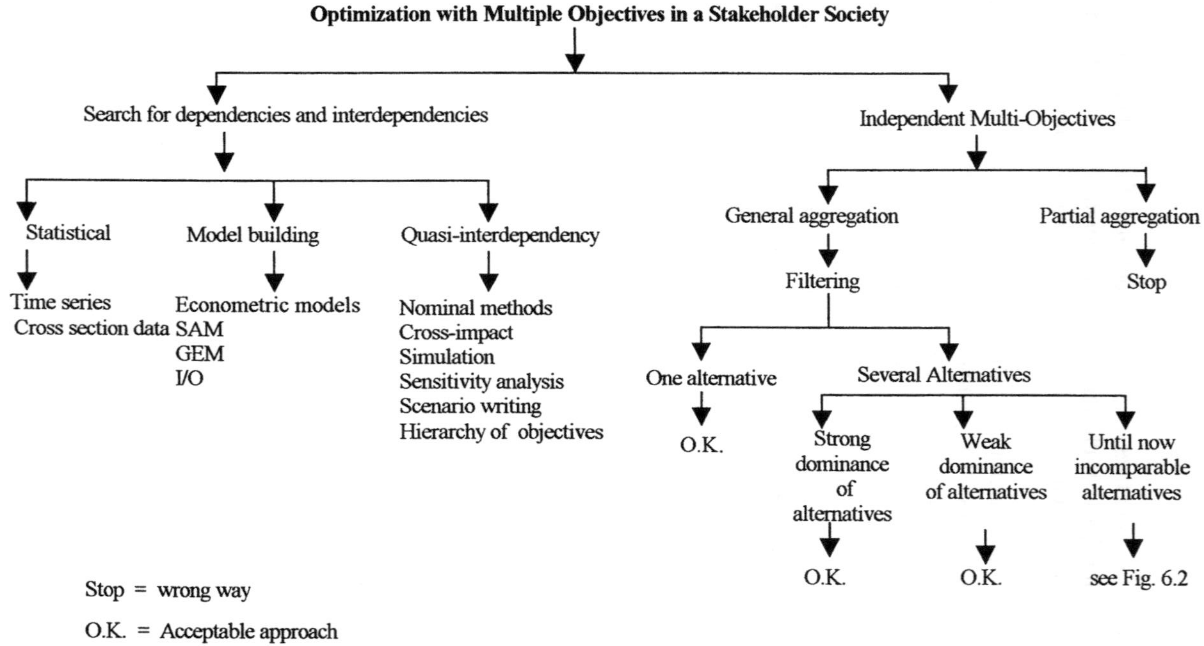

Fig.6.1. Dependency, Interdependency and Independency in Multiobjective Optimization

The problem with statistical relations, like *Time series*, resides in the fact that they are too much linked with the past. A mixture of time series and *Cross section data* may help. Too much linkage with the past forms also a problem for the interrelations in traditional *Econometric Models*, *Social Accounting Matrices* (SAM) and *General Equilibrium Models* (GEM). *Input-Output Analysis* is already a bit better, as production structures are mostly rigid for a limited time period. Two examples of input-output applications were given with multiple objectives, expressed in different units. One example was maximizing final demand and labor productivity at the same time. The other was introducing pollution abatement (see Part I, 2.7). Rather vague relations, which we call *Quasi-interdependency*, are found in nominal methods, cross impact, simulation, sensitivity analysis, scenario writing and in a hierarchy of objectives (see Part II). Automatically, besides dependent and interdependent multiple objectives, one has to speak of Independent Objectives. However, is independency well delimited? Perhaps all Objectives are at least of a quasi-interdependent type. To be operational, we shall continue to speak of Independent Objectives.

A *Hierarchical Objectives Structure* may exist under the form of a pyramid with subobjectives and sub-subobjectives descending with increasing specificity and decreasing permanency. The specificity finally boils down to measurability. It is a top down approach (see Part III, 1.6). In this way, the approach is general in the philosophy of the economists Walras and Leontief and not partial in that of Marshall and Keynes (see Part I, Chapters 1 and 2). Nowadays, in order to be operational, some methods of multiobjective decision making use partial aggregation (see Part III, 1.1). This behavior does not solve the problem of the optimal fulfillment of all objectives simultaneously.

The individual is striving for several objectives and if they are fulfilled in an optimal way, happiness has to be reached. This complicates optimization of several objectives simultaneously. However, it will be difficult to satisfy each objective. Consequently, we try to optimize each objective as much as possible and to satisfy every individual. The Green Parties in Europe maintain that economic theory cannot solve this problem of human happiness. They consider economics only as a science for minimizing costs. Therefore, we promulgate the idea of "New Economics", whereby all objectives are consider to the greatest extent possible in an optimal way.

In fixing the objectives, one has also to take into account not only every individual, but also all kinds of groups such as trade unions, employer organizations, consumer organizations and groups of ecologists or lobbies of all kinds. In addition, the state considers itself as the promoter of the general interest, ending up in Social Welfare or even in Social Well-being. All these agents may pursue different, even contradictory, objectives. Multiple objectives and multiple parties or partners represent an essential part of the stakeholder society we have to take into account.

Aiming to satisfy multiple objectives, we are faced with several potential solutions, projects, actions, which we simply call "*Alternatives*". Several attributes characterize these alternatives. An "*Attribute*", is a quantitatively well-defined common characteristic of each alternative, such as its economic, social, cultural or ecological significance. An "*Objective*" consists in the optimization (maximization or minimization) of an attribute. We avoid using the term "*Criteria*". Indeed, this term creates much confusion as being used for objectives as well as for attributes.

Quantities have to be well specified (see Part III, Chapter 1). First there is a problem of measurement. It makes no sense to carry around a theoretical notion of utility all the time. Therefore, all kinds of indicators of value have to be used. It is true that a problem of quality measurement exists. *Nominal scales*, i.e., the use of labels, like good, indifferent and bad, have to be avoided. Second, we prefer direct measurement, or at least alternative measurement. If this is not possible, exceptionally, persons interested in the issue will be asked to show their appreciation by giving a dimensionless cardinal number to the attribute per alternative. As human simultaneous perception is limited, this choice of cardinal numbers has to be very limited, for instance to three, but never more than around seven (see Part III, 1.5). Once the co-ordinates of all alternatives are determined, comparisons may start. A *Filtering Stage* may already drop certain alternatives not responding to certain restraints (see Part III, 3.2). If at that moment a single alternative remains, the problem is solved.

With several alternatives, one or some alternatives may dominate the others. If one alternative dominates all the others for all attributes, then this alternative is ranked first. An alternative, which is dominated by the others for all attributes, is ranked last (see Part III, 3.3). After filtering and domination, until now incomparable alternatives remain (see Fig.6.2).

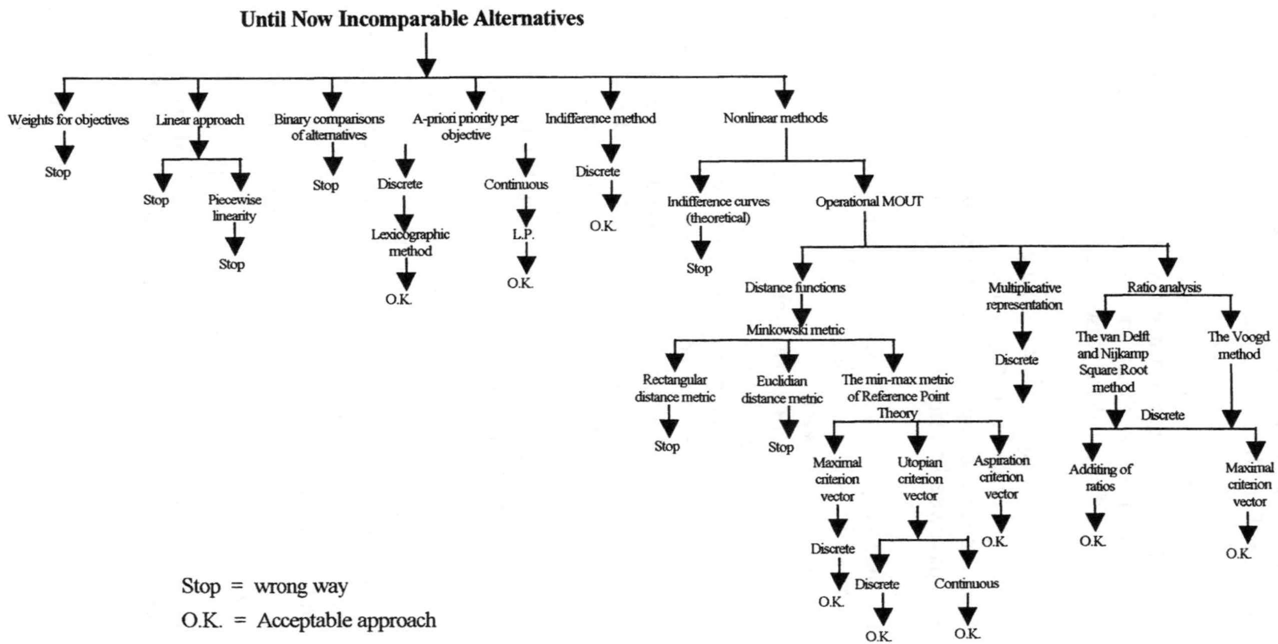

Fig. 6.2 Solutions for Until Now Incomparable Alternatives

If *Weights* are given to the several attributes a superobjective is created, which means mono-objectivity (see Part III, 2.4 and Part IV, 1.2). Moreover, weights favor alternatives which are extremely good in one or two objectives and disfavor in-between alternatives (see Part IV, 1.3). General agreement on weights between the stakeholders could be difficult, certainly when many stakeholders are involved. At that moment general agreement on the choice and number of objectives is already a sufficient point of discussion. For all these reasons, all methods using weights are not considered.

Linked to the use of weights, linear approaches are not welcome in multiobjective utility theory. Linearity infringes the economic law of decreasing marginal utility. Whereas, piecewise linearity forms a fair approximate of decreasing marginal utility curves, piecewise linearity is rejected for its use of weights (see Part III, 3.4).

It is true that at the supply side firms assume fixed linear production functions with constant returns to scale. At the end, however, they have to take into account decreasing marginal utility of final demand.

Also in macroeconomics, Input-Output works with fixed linear production functions. In the medium term or due to a sudden structural trend break, such as the energy crises in the past, in any case the production functions have to change.

If more than two alternatives are compared the comparison has to be *Multifarious* and not two by two with generalization afterwards. Indeed these binary comparisons could lead to contradictions or to reverse orders (see Part III, 2.5)

The possibility exists that the stakeholders fix in advance a priority order for the objectives realization. If the alternatives are known in a limited number (discrete), the choice is easily made (*Lexicographic method*). If the alternatives are generated during the process itself (the *Continuous case*), linear programming will bring the outcome (see Part III, 3.1).

In the discrete case after a hard filtering stage eliminating weak alternatives, the stakeholders may be indifferent in the choice of the alternatives. At that moment, the cost side will bring the outcome. Unanimity in indifference between the stakeholders may be difficult, certainly when many stakeholders and many objectives are involved (see Part III, Chapter 4).

The introduction of nonlinear methods is the next point to consider. Theoretically, the outcome is brought by analyzing indifference curves, but indifference curves are difficult to draw (see Part III, 3.4). Therefore, substitutes have to be found. Nonlinear *Multiple Objective Utility Theories* (MOUT) may come nearby indifference curve theory and ipso facto the economic law of decreasing marginal utility. Three kinds of nonlinear theories will be analyzed: *Distance Functions, Multiplicative Forms* and *Ratio Analysis*.

The *Minkowski metric* summarizes all the *Distance Functions*, bringing a good summary for the so-called *Goal Programming* (see Part IV, 1.7). On theoretical grounds, the *Rectangular distance metric* and the *Euclidean distance metric* are rejected, which is also confirmed by an application for the Euclidean distance metric (see Part IV, 1.12). For the same reasons all other distance functions are rejected with exception of the *Min-Max Metric*, which will form the basis for the *Reference Point Theory*. A reference point can be characterized in three ways: by a maximal criterion vector, a utopian criterion vector or by an aspiration criterion vector (see Part IV, 1.4).

The *Maximal Criterion Vector* is composed of the dominating co-ordinates of the alternatives per attribute. The Maximal Criterion Vector is a realistic approach. This is very important, when the alternatives are well defined, as for projects in Project Analysis and Project Planning. However, some call it only a satisficing result or bounded rationality, as it seems that the stakeholders are completely satisfied if the realistic reference point is reached. As the maximal criterion vector is linked to given alternatives, it is only applicable in the discrete case.

The *Utopian Criterion Vector* gives higher values to the co-ordinates of the reference point than the maximal criterion vector. This is understandable for instance for *Performance Management*, for student evaluation and for any performance in the private and the public sector. A desirable reference point as in the utopian criterion vector method is defendable. However, the reference point in question is determined in a subjective way. With the sky as limit the utopian criterion vector is applicable for discrete as well as for continuous cases.

The *Aspiration Criterion Vector* moderates the aspirations by choosing smaller co-ordinates than in the maximal criterion vector. Indeed stakeholders may be moderate in their expectations. Linked to the maximal criterion vector, the aspiration criterion vector is only limited to discrete cases.

The *Multiplicative Representation* is another nonlinear method for multiobjective optimization (see Part V). The proposed *n*-power form is not appropriated for transformation to a logarithmic form for at least two reasons. First, in the proposed formula an independent term makes intervention or correction by the stakeholders always possible. Second, all figures, applied in the formula and expressed in whatsoever unit, are reduced to a limited number of digits without changing the outcome. Raising to powers is always possible, but is only decided by unanimity or as the outcome of a *Delphi* exercise.

Is *Ratio Analysis* a separate method for multiple objectives optimization? Ratios of several kinds are suggested as a mechanical device. This mechanical approach takes the form of ratios with a common denominator. If weights or scores are determined with the previous methods, ratio analysis can be considered as a control method on these previous methods. However, ratio analysis seems to take a separate life. The question is then posited, if ratio analysis can act as a separate method for optimization with different objectives. Let us look at this enlargement (see part VI).

Several ratios for normalization can be thought of. We pick up some of them, which can be considered as representative. Anyway, one of the alternatives taken as a reference or the maximum value of an attribute observed in an alternative, are denominators for ratios, which are not appropriate for the purpose. The sum of all given values per attribute or the square root of the sum of their squares are denominators that are more appropriate. The corresponding ratios are then used twice: once for an adding method and once for the reference point method.

If weights favor alternatives which are extremely good in one or two objectives, on the contrary, the nonlinear methods afford an opportunity to alternatives with in-between positions concerning the objectives.

Figures 6.1 and 6.2 form a log on the methods for the multiple objectives researcher.

What are the opportunities for the *Continuous Case*, where the process of optimization generates a solution from a continuous set of alternatives? Here also the arguments against weights are maintained (see Part III, 2.4 and 3.4). With a *Complete A Priori Priority* between the objectives, consecutive linear programming is used (see Part III, 1.6 and 3.1). Otherwise, *Reference Point Theory* with a *Desirable Reference Point* is proposed for the continuous case (see Part IV, 1.10).

The very detailed description of nonlinear methods for multiobjective optimization may look quite theoretical. Therefore, the theory was tested with numerous applications. From these applications we came to the following conclusions.

First, *Reference Point* and *Multiplicative Representation* methods can be kept as full-fledged nonlinear methods for multiobjective optimization.

Second, *Ratio analysis* is more than a control method for the other methods. Ratio analysis may be considered as a separate and complete nonlinear method for *Multiobjective Optimization.*

INDEX

MIX
Papier aus verantwortungsvollen Quellen
Paper from responsible sources
FSC® C105338

If you have any concerns about our products,
you can contact us on
ProductSafety@springernature.com

In case Publisher is established outside the EU,
the EU authorized representative is:
Springer Nature Customer Service Center GmbH
Europaplatz 3, 69115 Heidelberg, Germany

Printed by Libri Plureos GmbH
in Hamburg, Germany